CLUSTER ION-SOLID INTERACTIONS

CLUSTER ION-SOLID INTERACTIONS

Theory, Simulation, and Experiment

Zinetula Insepov

Argonne National Laboratory
Illinois, USA

CRC Press
Taylor & Francis Group
Boca Raton London New York

CRC Press is an imprint of the
Taylor & Francis Group, an **informa** business

CRC Press
Taylor & Francis Group
6000 Broken Sound Parkway NW, Suite 300
Boca Raton, FL 33487-2742

First issued in paperback 2019

© 2016 by Taylor & Francis Group, LLC
CRC Press is an imprint of Taylor & Francis Group, an Informa business

No claim to original U.S. Government works

ISBN-13: 978-1-4398-7542-1 (hbk)
ISBN-13: 978-0-367-86661-7 (pbk)

Library of Congress Cataloging-in-Publication Data

Names: Insepov, Zinetula Z., author.
Title: Cluster ion-solid interactions : theory, simulation, and experiment / Zinetula Z.
    Insepov.
Description: Boca Raton, FL : CRC Press, Taylor & Francis Group, [2016] | "2016 |
    Includes bibliographical references and index.
Identifiers: LCCN 2015042110| ISBN 9781439875421 (hard cover) |
Subjects: LCSH: Ion bombardment. | Cluster theory (Nuclear physics) | Solids--Effect of
    radiation on.
Classification: LCC QC702.7.B65 I55 2016 | DDC 539.7--dc23
LC record available at http://lccn.loc.gov/2015042110

Visit the Taylor & Francis Web site at
http://www.taylorandfrancis.com

and the CRC Press Web site at
http://www.crcpress.com

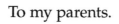

To my parents.

Contents

Foreword

I've known Professor Insepov since 2002, when we started working together in our Section of Computational Physics and Hydrodynamics of Energy Technology Division at Argonne National Laboratory (ANL). Later, when our group had moved to Purdue University, Dr. Insepov stayed in the Mathematics and Computer Science Division of ANL and worked there until he was offered a chief scientist position at a new, Western-style university in Astana, Kazakhstan. During that time, he was working as an adjunct professor at the School of Nuclear Engineering at Purdue University. There, he spent much effort in helping young students and young researchers to grow and become real scientists by giving excellent lectures and seminars and helping to supervise doctoral students.

This book is an enthusiastic presentation of the fundamental physical principles and mathematical tools that lie behind the new cluster ion beam technology mostly developed in Japan within the last two decades. The book explains the basic principles and fundamentals of cluster nucleation, formation, and growth in various media, including adsorption layers on the surfaces, in rarefied and dense gases, as well as on various applications of cluster ion beams for surface cleaning and etching and implantation. By explaining the process, it builds a bridge between the fundamental concepts of thermodynamics and physical kinetics with the new physical and engineering cluster applications.

Clusters and cluster ions are among the most abundant media in the universe. Since their introduction into thermodynamics by Boltzmann in 1903, there have been numerous papers and books published discussing the idea that density fluctuations in gases or liquids could be stable physical formations. Eventually, clusters were considered assemblies of neutral gas atoms bounded together with weak van der Waals forces. This idea was important and fruitful for validating Classical Nucleation Theory (CNT), which is still an area requiring strong experimental justification. Therefore, the idea of the cluster encourages experimentalists to search for new effects and understanding in this important engineering applications area.

My own experience with clusters started from understanding that the development of new extreme ultraviolet lithography (EUV) technology initiated by the semiconductor industry could be severely affected by cluster ions bombarding the light-focusing mirrors. These clusters are formed in either a laser-ablated plasma or in a gas discharge plasma used to generate 13.5 nm photons for future EUV nanolithography.

Another area that has benefited from clusters is Tokamak physics and engineering, where helium clusters that could be formed in liquid lithium are proposed to be used as the plasma-facing component in future thermonuclear reactors. Accumulation of plasma particles in liquid surfaces and their ejection could lead to a disruption in plasma operation, and therefore simulation and modeling by atomistic methods is important to shed light on the mechanisms of interaction of plasma with surfaces.

The study of radiation defects in nuclear materials is a third area that has benefited from the idea of clusters: interstitial and vacancies form clusters moving fast in metals and the physics of such clusters could be explained by standard cluster physics.

In addition to summarizing the fundamentals underlying cluster ion beam technology, this book is also a unique tribute to the many scientists who were involved in developing cluster studies and applications, most of whom are cited in this book. Although the emphasis of this work is on atomic clusters, it contains much information that will be of interest to those outside the field and to students of nucleation thermodynamics—indeed to anyone with a fascination with the world of clusters.

The author has selected well over 50 of his own papers as the key subjects of his text. Although these represent only a small sample of the world of cluster-related areas, they amply illustrate the importance of this field of science to humankind and the way the field has evolved. I believe that the author can be confident that there will be many grateful readers who will gain a broader perspective of the disciplines of cluster science and applications as a result of his efforts.

Ahmed Hassanein
Purdue University

Preface

The idea to write a book about clusters was conceived by my editor, who was visiting Argonne National Laboratory (USA) in 2011. I was working there on a completely different project: developing a prospective nuclear fuel. Soon I realized that radiation defects in nuclear fuels have a benefit in that they form clusters of defects, similar to gas clusters. The cluster concept has been fruitful in fields that are far from traditional cluster applications, so it makes sense to have a general treatment for undergraduate and postgraduate students of physical and technical specialties.

The main goal of this book, therefore, is to provide the reader with an overview of various concepts in cluster physics and related topics in physics, and is based on lecture courses given by the author to undergraduate and graduate students at Kyoto University (Japan), to nuclear engineering students at Purdue University (USA), and to biology and molecular physics students at the Moscow Institute of Physics and Technology (MIPT) and the Moscow Engineering Physics Institute (MEPhI), both in Russia. The main results included in this book were obtained during my research and development work on cluster ion beam applications while I was working at the Ion Beam Engineering Laboratory of Kyoto University under the guidance of Professor Isao Yamada, who is my coauthor of almost all the research papers in this field.

For many years, I held a unique educator and research position, while being a theoretician at a large experimental laboratory oriented toward industrial applications. Therefore, the main method of writing this book has been to start with introducing the basic principles of statistical physics and thermodynamics and then move to practical applications, which are followed with experimental justifications and practical implementations. The book starts with a description of Classical Nucleation Theory and shows the drawbacks of this theory, demanding accurate modeling and simulations in order to justify theoretical approaches and simplifications.*

I wish to acknowledge several people who were most influential to me during my work on this subject for many years: Professor Genri Norman (MIPT) significantly encouraged me to start working in the area of modeling and simulation in cluster physics since I was an undergraduate student in Moscow, Russia; Professor Hellmut Haberland (Freiburg University, Germany) showed me how to combine simulation with experiment, thus making realistic theory and simulation; Professor Isao Yamada (Kyoto University) gave me a great example of an innovative approach to cluster ion beam and applications; Professor Ahmed Hassanein was my mentor during the US work at Argonne and now at Purdue. I am indebted to all these scientists and educators. I especially acknowledge my long-term colleagues Dr. James Norem and Dr. Jeffrey Rest (both from Argonne) for their continuing friendship and fruitful collaboration.

Zinetula (Zeke) Insepov
Darien, Illinois

---

* It should be noted that a central result showcased in this book is the modeling and simulation prediction of so-called lateral sputtering, obtained in 1992 and published in 1993. At that time, the new phenomenon was unusual since ions bombarding the surface were ejecting substrate atoms according to a classical cosine law, where the atoms were mostly ejected in a backward direction to the incoming ion.

Acknowledgments

The author is indebted to Professor Isao Yamada and the staff of the Ion Beam Engineering Laboratory at the Graduate School of Engineering of Kyoto University, Japan, for their dedication and hard work over more than 10 years of collaboration. They have made great efforts in performing experimental discoveries and giving me an excellent opportunity to develop my theory and simulation work, as well as verifying the results of numerous computer simulations.

Thanks are also due to Professor G. E. Norman and to his research team at the Joint Institute for High Temperatures at the Russian Academy of Sciences, and to Dr. Alexander Yu. Valuev, Dr. Victor Yu. Podlipchuk, Dr. Sergey V. Zheludkov, Dr. Essey M. Karatayev, and Dr. Aktorgyn M. Zhandadamova for their invaluable help and discussions over the years. These stimulating conversations have indeed helped in maintaining the motivation in many large and long-term studies.

I am also thankful to my former colleagues Professor Marek Sosnowski, Professor Jiro Matsuo, Professor Gikan Takaoka, Dr. Rafael Manory, Dr. David Swanson, and Allen Kirkpatrick at Epion Corp (now Exogenesis Corp), for their encouragement, approval, and tolerance during the many years of collaboration.

I am greatly indebted to my former students and colleagues Dr. Daisuke Takeuchi, Dr. Makoto Akizuki, Dr. Takaaki Aoki, Dr. Toshio Seki, Dr. Teruyuki Kitagawa, and Professor Noriaki Toyoda, who helped me with discussions on specific topics as well as with understanding their experimental results.

Acknowledgments

Author

Zinetula Insepov was a visiting professor and chair of new materials research at the Graduate School of Electrical Engineering, Kyoto University, Japan, from 1992 to 2000. For the next three years, he worked as a research and development manager of a private Japanese electronics company, Epion Japan, which specialized in cluster physics and technology. In 2003, he moved to the United States, where he started working at the Energy Technology Division and later moved to the Mathematics and Computer Science Division of Argonne National Laboratory. In 2013, he was appointed chief scientist of the Research and Innovation System at Nazarbayev University in Kazakhstan, the first Western-type university in Central Asia. In 2012, he was selected to be a founding faculty fellow and professor of the MIT/Skolkovo initiative in Russia. He continues teaching and research work as an adjunct professor at the School of Nuclear Engineering of Purdue University. He has developed cluster ion beam interaction simulation programs based on molecular dynamics and Monte Carlo methods. He also predicted a new lateral sputtering phenomenon that is one of the driving forces behind the efficient atomistic smoothening mechanism of surfaces irradiated by a large gas cluster ions. His principal research over the past 20 years has addressed the fundamental physics of ion beam materials processing, including very-low-energy ion–solid interactions.

CHAPTER **1**

Introduction

This chapter gives a detailed introduction into the basic concepts
of vapor-to-liquid phase transformations, such as physical clusters as the nuclei of the new phase, the formation of such clusters
in gases, the kinetics of cluster formations in dense gases and on
solid surfaces, the interaction of gases with solid surfaces, and,
finally, molecular dynamics (MD), a powerful computer simulation
method that is the main method of studying the thermodynamics
and kinetics of cluster formations.

Section 1.1 describes physical clusters as aggregates of small
numbers of atoms or molecules that are bound together via weak
van der Waals forces. The size distribution of clusters is given by
a Gibbs's formula. Section 1.2 discusses cluster formations in gases
and presents the Classical Nucleation Theory (CNT), which is based
on the concept of liquid nuclei. This section shows that the CNT
predicts the supersaturation level in diffusion chambers and can
be used for calculating the condensation rate in a supersonic jet; for
example, Section 1.2 also studies condensation behind the shock
wave (SW) front in the presence of chemical reactions. A SW is an
abrupt increase of pressure or density that propagates through the
media, such as gas. A SW saturates the gas with a relatively high
density of metal atoms. A simplified condensation model is developed where the main interaction processes between the species are
monomer–monomer and monomer–cluster collisions. In Section 1.3,
an analysis of the kinetics of cluster formation in dense gases is given.

Section 1.4 discusses spinodal decomposition in dense gases that
are characterized by periodic density oscillations in time. It means
that all wavelengths will be damped except for one. According to
this theory, a temporal and spatial variation of the density is defined.

1.1   Introduction

The thermodynamic and kinetic theories of phase transitions were
developed in the classical works of Gibbs [1], Volmer and Weber [2,3],

Becker and Döring [4], Zel'dovich [5], Frenkel [6], Turnbull and Fisher [7], Hirth and Pound [8], and Zettlemoyer [9], and continued in the recent developments in the simulation and modeling of nucleation phenomena by Stein and coworkers [10], Burton [11], Abraham [12], and Smirnov et al. [13], and is still being developed, since many features of nucleation at high-level supersaturation are not yet fully understood.

The fundamental formulas of this theory can still be characterized by large uncertainties and are sometimes contradicting. The most difficult problem is building a sound theoretical basis of the nucleation processes of a new phase in dense media [7].

According to Gibbs' ideas [1], an increase of the supersaturation level would lead to a rapid decrease of the critical radius of the nucleus that would be capable of growing to a new phase. Estimates of this critical radius for contemporary experimental conditions show that this parameter can be as small as of the order of a unit. In this case, obviously, the CNT becomes inapplicable due to the fundamental limitations of the latter [6,8]. Furthermore, the discussions are still active in the literature on the contribution of the translational degrees of freedom of a small critical nucleus (cluster) into the free energy of the new phase [11].

At the same time, this theory has been used widely in numerous technical devices and equipment in aerodynamics, vacuum techniques, microelectronics, nuclear materials, and fuel developments and applications on industrial scales [14–19].

Atomistic simulation methods, such as Monte Carlo (MC) and MD, have significantly improved our understanding of various equilibrium properties of different systems [12]. However, the capacity of these methods is much wider and, with a little modification they can, for example, be applied to more complicated physical processes of a new phase formation, and it also allows one to obtain elemental mechanisms of the new phase formation [13]. Specifically, the above statement is relevant to MD, which, according to its nature, is an exact modeling tool of both the dynamics and the kinetics processes in the matter.

The atomistic methods (MC, MD) therefore are capable of first verifying the exact theoretical models that can be solved analytically and then studying the areas of experimental parameters that are unreachable at the current state of the art and hence they can predict the properties at conditions that cannot be retrieved easily.

The goal of this book is to review and summarize existing theoretical and simulation models of cluster formation and discuss applications of the theory and simulation to various technological processes and equipment apparatus.

Another goal is to develop new theoretical and simulation models involving clusters in such areas as accelerator physics and radiation defect formation in nuclear fuels. The development of future high-gradient teravolt linac systems is significantly slowed down by the high-voltage breakdown that occurs when a dense metal plasma forms close to the surface and a Coulomb explosion mechanism of clusters plays an important role in the evolution of the breakdown. At a high electric gradient the rough surface containing sharp corners and tips starts emitting large chunks and small clusters of surface material that eventually breaks down the vacuum and leads to the power deposition to the wall. Simulation tools have been developed that are capable of eliminating the uncertainties of previous theories and help us to understand the whole phenomenon.

The cluster concept is very important for studying fundamental questions of fission gas bubble nucleation and growth that becomes very important for the development of low enriched nuclear fuels.

The behavior of gas atoms and defects in metal fuels can be understood by using the same theoretical methods that were developed for typical gas–liquid phase transitions.

Various physical processes and interaction of ions with the surfaces were studied by atomistic simulation methods that have currently become one of the major methods for understanding and making the perditions. For example, the phenomenon of lateral sputtering was predicted by MD simulation in 1992 and later was confirmed by experiment. The cluster formation concept was shown to be fruitful in many research areas including phase transformations, radiation defect formation and fuel swelling, accelerator physics, and microelectronics. New kinetic processes including defect formation and new phase formation in various physical applications, such as vapor-liquid, melt-crystal, thin film growth, and surface erosion by energetic ions were discussed in this part of the book.

To address the above-mentioned tasks, the following tools were developed in this work:

- A series of simulation codes and new molecular models of phase transitions for unsaturated and saturated dense vapor

- Cluster formation in the adsorption layer on the surface
- Sticking of gases to the surfaces at elevated velocities
- Erosion of surfaces by energetic cluster ions
- Modification of the MD method for studying kinetics at phase transformations

1.2   Clusters: Nuclei of New Phase

Physical clusters are aggregates of small numbers of atoms or molecules that are hold together via weak van der Waals (vdW) dispersion forces. Since the vdW adhesion forces are very weak, the sizes of such atomic aggregates are usually much larger than an atomic size. These physical clusters are quite different from chemical clusters that are bond together via strong chemical bonds and which are closer to chemical molecules from the prospective of many properties.

The physical clusters are the fundamental building blocks of the new phase in all vapor–liquid phase transitions of the first kind in which the initial system (vapor) is abruptly moved out from an equilibrium state above the coexistence line (binodal) into an area in the phase diagram located between the binodal and spinodal (Figure 1.1).

The concept of a physical cluster and the corresponding term were apparently introduced by Ya. I. Frenkel, a famous Russian physicist, in his classical monograph published in 1946 [6]. The author termed them *heterogeneous fluctuations* and developed a general thermodynamic theory of the formation of such clusters.

The distribution of clusters of a new phase along the coexistence curve is stable if the chemical potential of the new (daughter) phase μ_B is lower than that of the old (mother) phase μ_A. It is given by Gibbs's formula:

$$N_g = C \exp\left(\frac{-\Delta G}{k_B T}\right),$$

(1.1)

where: $C = N_A + N_B$
where:

N_A, N_B are the total number of old (A) and new (B) phase atoms/molecules per unit volume, by assuming that $N_B \ll N_A$

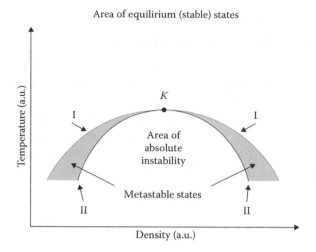

Figure 1.1 Schematic representation of characteristic areas for the case of vapor–liquid phase transitions of the first kind: K—critical point, I—binodal, II—spinodal.

ΔG is the free energy of the cluster containing g-atoms (molecules)

T is the temperature of the old (mother) phase

Turnbull and Fisher applied the kinetic theory of absolute reaction rates to derive an expression for the nucleation rate and Equation 1.1 for liquid–liquid and solid–solid phase transitions in condensed media [7].

Recently it has become clear that the concept of the cluster state of matter is an important one, as it is an intermediate substance between free atoms (molecules) on one side and a macroscopic phase on the other side. Several theories were developed that assumed the universality of the clustered structure of gaseous, liquid, and even solid matter [12,20–22].

There exists a vast literature on cluster physics and chemistry such as cluster generation, stability and structure, and electrical, magnetic, and optical properties. These studies were widely represented at the International Conferences in Small Particles and Inorganic Clusters (see, e.g., in the *Proceedings of the Conferences on the Physics and Chemistry of Finite Systems: From Clusters to Crystals* [23–25], and the Springer Series in cluster physics [26–28]).

1.3   Cluster Formation in Gases

1.3.1   Classical Nucleation Theory (CNT)

CNT for a supersaturated vapor that was developed by the work of Volmer and Weber [2,3], Becker and Döring [4], Zel'dovich [5], and Frenkel [6] gave a significant understanding of the behavior at phase transitions of the first kind in a rarefied vapor.

In this theory, the basic idea rests on the concept of a liquid nucleus where the free energy ΔG of the critical droplet r^* is represented as a sum of two parts: one is for the surface part and the second one is for the bulk part [7–9]:

$$\Delta G = \left(\mu_l^\infty - \mu_g\right)n + 4\pi r^{*2}\sigma ft, \tag{1.2}$$

where:

μ_l^∞, μ_g are the chemical potentials of the liquid and gas phases, respectively

n is the number of atoms (molecules) in the nucleus

σ is the surface tension coefficient for the interface between the liquid and gas phases

The stationary rate of condensation in such conditions is

$$J_0 = \alpha_c 4\pi r^{*2} \frac{P}{\sqrt{2\pi m k_B T}} \exp\left\{-\frac{4\pi\sigma\left(r^*\right)}{3k_B T}\right\}, \tag{1.3}$$

where:

α_c is the probability of a single atom (molecule) (a monomer) condensing at the surface of a critical cluster in a binary collision between the cluster and a gas atom

P is the gas pressure

m is the atomic mass

T is the ambient temperature

The critical radius of the nucleus r^* is defined in these formulas by maximizing the free energy of droplet formation and is equal to

$$r^* = \frac{2m\sigma}{\rho_l k_B T \ln S}, \tag{1.4}$$

where:

ρ_l is the liquid phase density

$S = N_1/N_{eq}$ is the supersaturation rate in the system

N_1 and N_{eq} are the current and equilibrium densities of atoms (molecules) in the gas

Careful comparison of the CNT with numerous experimental data shows good applicability of the theory to the results obtained in experiments with diffuse cameras. A qualitative comparison was also obtained for the condensation of a SF_6 molecular gas in supersonic nozzle expansion experiments in which argon was used as a carrier gas [10].

However, the experimentally measured nucleation rates J_{exp} were drastically different from those calculated by Equation 1.3, by a factor $\sim 10^2$–10^5 [29,30].

In [29], the authors determined that using the values of ΔG (r^*) calculated by the Monte Carlo simulation method allowed them to qualitatively analyze the experimentally observed nucleation rates. The main reason for such an interpretation is that the authors related to the small number of atoms in the critical cluster. Under the experimental conditions in [29], the critical nucleus contained only from 9 to 18 molecules, which makes the liquid droplet model inapplicable to this size.

A full analogy exists between the work in [29] and the results measured at a supersonic gas expansion [31,32] and behind the SW front [33]. The critical nucleus for the latter two cases contains only a few molecules. Therefore, a possible solution in this situation would be to use real properties of the clusters, such as formation enthalpy, heat capacity, and cluster formation entropy, that can be measured experimentally or calculated numerically.

In [34], the agreement between the theory and experiment was obtained by using a surface tension coefficient that depended on the cluster radius. Bedanov [35,36] put forth efforts to directly calculate the stationary nucleation rate J_0 via the direct simulation of the cluster evaporation process. The following formula was used:

$$J_0 = \frac{N_1 c_1}{\left(1 + \sum_{g=2}^{n^*-1} \prod_{n=2}^{g} e_n/c_n\right)}, \tag{1.5}$$

where:
    c_n and e_n are the rates of condensation and evaporation for a
    cluster of size n

    c_n $= \alpha_n N_1 S_n (k_B T/2\pi m)^{1/2}$

Here, α_n is the sticking coefficient; S_n is the area of a cluster of n
atoms (molecules). e_n data were obtained by an MD method. The α_n
were set to 1. As a result, a good comparison of J_0 with experiment
was obtained. The sticking coefficients α_n of a gas particle to a cluster containing $n = 2$–100 atoms were calculated by an MD method
[37]. The probability of dimer formation was calculated via a three-
body approach. The obtained values of α_n were substituted into
Equation 1.5 and the nucleation rate J_0 was calculated. This analysis showed that the nucleation rate is changed by several orders of
magnitude. The largest contribution to J_0 was made by the dimer
formation probability α_2.

In principle, having knowledge about all probabilities of formation and decay of small clusters lets one finally solve the problem
of applicability of CNT for condensation at very high supersaturation rates that are distinctive for supersonic flows, SWs, and Wilson
chambers [38].

1.3.2   Probabilities of elemental reactions in the cluster formation on gas

A rigorous theoretical analysis of the probabilities of cluster formation and decay of an arbitrary size is an almost unsolvable task
due to a known three-body problem in mechanics. In [31,32,39,40]
the rates of dimer formation and decay were calculated. It was
assumed that dimers can be formed via a three-body collision having a rate constant k_f and it decays with a rate constant of k_r in a
binary collision:

$$A + A + M \underset{k_r'}{\overset{k_f'}{\rightleftharpoons}} A_2 + M, \tag{1.6}$$

where:
    A, A_2, M correspond to monomers, dimers, and a noncondensing
gas atom

    A gas dynamics three-body formula was used to calculate k_f.
A statistical model of a three-body recombination developed for
hydrogen atoms was adapted to clusters in [41].

Calculations were performed for Ar and CO_2. Theoretical values of dimerization constants k_f were deviated from the experimentally measured in [41] to an order of magnitude for Ar and for five orders of magnitude for CO_2.

The main difficulty is calculating the sticking coefficient α_n that immediately defines the rate constant of the cluster formation c_n via a hydrodynamic formula. Usually, these coefficients are obtained by comparing experimental data with the calculated rate constants [18,42].

Gordiets et al. [15] found α_n from the time dependence of the average radius of lead clusters measured in [42]. The authors obtained good agreement with experiment for the average condensation time τ_k by assuming that the sticking coefficient is in the range $\alpha \approx 0.4–0.7$. These authors have also obtained the sticking coefficient using the cluster growth rate constant and have derived the sticking coefficient to be equal to $\alpha = 0.33$.

In [43,44], the coefficients of sticking of atoms to clusters were calculated by comparing the numeric solution of a set of kinetic equations for the concentrations of clusters of different sizes with experiments. Neglecting the evaporation of clusters, the authors of [44] obtained the following average values: $\alpha_2 = 3.3 \times 10^{-2}$, $\alpha_3 = 3.6 \times 10^{-2}$, and $\alpha_4 = 3.8 \times 10^{-2}$.

Another rate constant characteristic for the reaction (1.6) is the decay constant k_r of a dimer A2, at a collision of this dimer with any particle in the system M. In [31,32,43] this rate constant was derived via an equilibrium reaction rate constant K_{eq} for the reaction (Equation 1.5):

$$k_r' = \frac{k_f'}{K_{eq}}$$

$$K_{eq} = \left(\frac{h^2}{\pi m k_B T} \right)^{3/2} Z_{2,\text{int}},$$

where:

  h is Plank's constant
  m is the atomic mass
  $Z_{2,\text{int}}$ is the internal configuration integral of a dimer

To calculate the configuration integral (statistical sum) of the dimer, one can use various classical and/or quantum methods. For example, in [31,44] the quantum statistics is replaced by a classical, the Schrodinger equation is solved numerically in [32], and experimental data are used in [45].

The inelastic processes during collisions of a particle M with a cluster A_n, can be represented as the following two-level schematic:

$$M + A_n \underset{k_s}{\overset{k_e}{\rightleftarrows}} M + A_n^*. \tag{1.7}$$

Here, A_n^* corresponds to a n-cluster that has excess internal energy.

1.3.3   Calculation of cluster formation in gas via molecular dynamics

Brady, Doll, and Thompson [46–48] showed that the rate constant for evaporation of a single atom from an excited cluster can be calculated by using an MD method (a classical trajectory method by the author's terminology). A vibrationally or rotationally excited cluster was prepared by the following method. A cluster consisting of $n + 1$ atoms interacting via a Lennard-Jones potential was placed at the origin of the coordinate system and hit by a test atom, with a thermal velocity $v = (3k_B T/m)^{1/2}$ and a certain collision parameter. The thermal velocities of the cluster atoms were set to zero for simplification. Several outcomes were observed as a result of such collisions: excitation of the cluster (the cluster gain the internal energy); fragmentation of the cluster atoms to smaller species; and elastic reflection of the atom from the cluster. To obtain the probability of excitation, only those trajectories were considered that would create an excited configuration consisting of $n + 1$ atoms with an additional internal energy, and a canonic ensemble of temperature T was formed from such collisions. The obtained probability was averaged over the collision parameter and the initial coordinates.

Each ensemble contained at least 1000 initial atomic configurations. Starting from some time instant t_0, when the probe atom approaches the cluster to a minimal distance, the number of survived clusters $dN/N(\tau)$ was obtained, that is, those clusters that did not emit a single atom after the collision, within the time interval of $\tau = t - t_0$. The obtained results were put to a plot of $dN/N(\tau) = f(t)$ and a characteristic cluster decay time was obtained as a slope of this plot.

One of the main results was confirmation that the monomolecular decay of a cluster consisting of $n+1$ atoms follows the theory of Rice-Ramsperger-Kassel (RRK) [49].

Chekmarev and Lyu calculated the evaporation rates of clusters using conventional and stochastic MD methods [50,51]. The calculations were performed within microcanonical (with constant total mechanical energy, $E=$const) and canonical (with a constant temperature) MD ensembles. The authors demonstrated that evaporation from the cluster can be explained adequately using the RRK theory. The parameter E_d was obtained 20%–30% below that calculated by Gadiyak and coworkers in [52,53].

Another new application of atomistic simulation to kinetic theory of cluster growth is MD calculation of the sticking coefficients of atoms to clusters. In [54,55] a probability of dimer formation in a ternary collision (simultaneous collision of three particles) was calculated by MD. The trajectories of colliding particles were analyzed, and those that were relevant for the three-body collisions were selected. If an internal energy of any of the atomic pairs became negative, the collision was counted as forming a dimer cluster. The probability of dimer formation for the temperature interval of $k_B T/\varepsilon = 0.85$–2.2 was obtained to be in the range of 0.05–1. In [55] the authors added that before the stable complex resulted in the collision, the colliding particle can form a metastable complex.

The most comprehensive calculation of the sticking of atoms to clusters was conducted by Bedanov and by Napari et al. [37,39]. The simulations explicitly included the cluster and the ambient gas. The projectile hit the cluster at different impact parameters and if the total energy in the system including both the projectile and the cluster became negative and less than 0.1ε, where ε is the binding energy of an atom in the cluster, then the sticking process was considered to be successful. The obtained values of the sticking coefficient were in the range of 0.1–0.9 for the temperatures $k_B T/\varepsilon = 0.01$–0.7 and the cluster size $n=47$. In the same paper, the author calculated and compared a sticking coefficient of an atom to a flat solid surface. The obtained coefficient was in the range of $\alpha_\infty = 0.44$–0.68, at temperatures close to a critical of the Lennard-Jones system.

Finally, the evaporation rate of small Lennard-Jones clusters was conducted in a microcanonical ensemble and the obtained result was compared to the RRK model. The calculation method was identical to the one previously used by Brady, Doll, and Thompson [46–48]. The results were obtained for cluster sizes $n=7$, 12–14. The average lifetime of the cluster was strongly dependent on the total cluster energy, which changed from 10 to 100 ns when the total energy declined. The authors found that a simplified RRK theory correctly describes the dependence of the evaporation constants on the total energy.

1.3.4 Condensation behind the shock wave (SW) front in the presence of chemical reactions

When an SW propagates through a gas containing additives of easily volatile metal–organic complexes, the SW instantly super-saturates the gas with a relatively high-density vapor of metal atoms. Contemporary techniques of SW registration are capable of registering and studying condensation in the vapor and station-ary states and controlled conditions. At the same time, this method can generate a very high supersaturation level, when the critical nucleus of the new phase is sufficiently small, according to CNT. All this makes this method attractive for understanding the nucle-ation phenomena behind the SW front. The condensation kinetics of the supersaturated gases behind the SW front was studied in the papers of Krestinin et al. [56], Gordiets et al. [15], Zaslonko et al. [33], and Bauer et al. [42].

The main specific of the condensation behind the SW front dur-ing the decomposition of carbonyls is the presence of a radical mol-ecule Me $(CO)_n$, where Me stands for the metal atom. Since such molecules have a high reactivity, they participate in chemical reac-tions and thus provide for the high dimer concentrations Me_2 [33].

The authors of [33] introduced a characteristic condensation time of Fe vapor for the case of iron pentacarbonyl $Fe(CO)_5$ decom-position in the SW. The carrier gas was noncondensing Ar. A sim-plified condensation model was used where the main processes were of the monomer–monomer and monomer–cluster types of collisions. It was assumed that iron atoms recombine with each other with the rate typical for three-body collisions and all further stages of growth occur by binary collisions. A maximum cluster growth rate was calculated and therefore the decay of the clusters was neglected. Under such approximations a time evolution of the iron concentration was obtained as

$$[\text{Fe}] = [\text{Fe}]_0 \exp\left(\frac{-t^4}{t_{\text{char}}^4}\right)$$

$$t_{\text{char}} = \left(\frac{108}{k^3 k_r [\text{Ar}][\text{Fe}]^4} \right) \qquad (1.8)$$

where:

    k is the rate constant of binary collisions Fe + Fe

    k_r is the constant of recombination of atoms

    $[\text{Fe}]_0$ is the initial concentration of iron atoms

By using the typical parameters for the experiment [33], the authors have shown that the typical characteristic condensation time for the iron vapor is an order of magnitude higher than the experimental value obtained in their work [33]. The obtained discrepancy was explained by a new chemical model of condensation according to which the cluster formation can obey the following reactions:

$$\text{Fe} + \text{FeCO} \rightarrow \text{Fe}_2 + \text{CO},$$
$$\text{Fe}_n + \text{FeCO} \rightarrow \text{Fe}_{n+1} + \text{CO}. \qquad (1.9)$$

In later work [56], the same authors developed the model further by including coagulation in the model iron cluster. The result obtained in [56] could satisfactorily describe the initial stage of the iron concentration time dependence and showed better comparison of the characteristic condensation time with experiment than that without reactions Equation 1.9.

1.4 Cluster Formation Kinetics in Dense Gases

CNT ceases to be applicable in systems where an atomic (molecular) density n approaches the reciprocal atomic volume v^{-1}:

$$nv \sim 1. \qquad (1.10)$$

However, well before expression Equation 1.10 approaches the kinetic equations for the cluster concentrations that use binary collisions, cross sections become meaningless. With the density rise, the atomic motion begins to resemble a Brownian motion, rather than straight lines and sudden collisions typical for rarefied gases, as is mentioned in [12]. So the kinetics of cluster formation will be diffusion limited. However, the diffusion constants are not included in the formulas for the critical radius.

CNT has a similar deficiency when a condensing gas represents a small additive to a carrier (noncondensing) gas with a high density.

It is well known [12,20] that the evolution of a condensing system depends on the initial state where the system was put. If the system initially resided in the equilibrium part of the n–T phase diagram, and suddenly transformed into an area between binodal and spinodal (Figure 1.1), then the cluster formation or nucleation is initiated. However, if the system is placed into the area below the spinodal, that is, into the area of absolute instability, then the system undergoes spinodal decomposition. Both of these mechanisms were investigated by atomistic computer simulation methods.

1.4.1   Molecular dynamics simulation of kinetics of cluster formation in dense gases

Brickmann and coworkers [57,58] conducted direct calculations of cluster formations in dense Ar and Xe gases. Gas atoms were placed into a basic cell with periodic boundary conditions. Interactions between the atoms were modeled by a Lennard-Jones potential [57] and Axilrod–Teller and exchanged three-body potentials [58].

Calculations were provided at a fixed temperature of 120 K and atomic densities $n\sigma^3 = 0.0553$ and 0.088, which corresponded to 1.4×10^{21} and 1.27×10^{20} cm^{-3} for the argon and xenon gases.

The following criterion was used for cluster definition: if the distance between two atoms was smaller than 2σ, where σ is the atomic diameter, then the two atoms were considered as bound into a cluster. In addition to this condition, which can be named a "geometric criterion," the authors put an additional constraint that the above simple criterion could not break within one vibration period of the dimer.

For better statistics, the results obtained by using different initial states were averaged. However, since each realization was time consuming, sufficient sampling was not provided. A relative number of clusters with different sizes and an average lifetime of clusters were calculated.

In [57], the authors have shown that using a simple pair-additive Lennard-Jones potential to describe the interactions between the atoms in the cluster leads to a monotonic dependence of the cluster binding energy on the cluster size. In the thermodynamic equilibrium region above the binodal, the authors obtained the maximal cluster size $n \sim 10$. Adding a three-body Axilrod–Teller potential to the interaction and exchange potential did not qualitatively change the obtained cluster size spectrum. However, this increased the number of clusters with the sizes $n > 6$ and lead to observations of larger clusters with sizes $n = 19, 21, 23$. If the temperature in the system increases, the role of the three-body interactions decreases.

The occurrence of relatively large clusters found in a simulation [57–59] of a thermodynamically stable gas above the binodal is surprising since the Frenkel theory of heterogeneous fluctuations predicts an exponentially small number of large clusters, with the sizes larger than $n \geq 5$. Some reasons for such a discrepancy could be a small-sized simulation system and a simplified criterion of the cluster identification procedure.

1.4.2   Spinodal decomposition in dense gases

If an equilibrium gas is suddenly placed into a region below the spinodal, that is, it arrives in an absolutely instable region, it starts transforming according to the spinodal decomposition behavior [60,61], which is characterized by periodic density oscillations in time and space in which all wavelengths will be damped except for one, which would be significantly magnified. This wavelength is called the critical wavelength and it is defined by the thermodynamic parameters of the system.

The theory of this phenomenon is based on a generalized diffusion law and was given by Cahn and Hilliard [62]. According to this theory, a temporal and spatial variation of the density can be defined from the equation

$$\frac{\partial \rho(\vec{r},t)}{\partial t} = M\nabla^2 \left[-K_\rho \nabla^2 \rho(\vec{r},t) + \frac{\partial f(\rho)}{\partial \rho} \right], \tag{1.11}$$

where:

$\rho(\vec{r}, t)$	is the local density of particles at a position $\vec{r}$ at a time instant t
K_ρ	is the coefficient in the expansion of the system free energy in the vicinity of the position $\vec{r}$
M	the mobility of the particles
$f(\rho)$	is the volume density of the free energy

Assuming the deviation of the density at an arbitrary location from the average value sufficiently small, that is, $\rho = \rho_0 + \Delta\rho$, the free energy density can be expanded into a series over $\Delta\rho$, keeping the terms up to the second order $\frac{1}{2}(\delta^2 f/\delta\rho^2)\Delta\rho^2$. Further, by using the Fourier transformation, Equation 1.11 can be linearized and solved. The solution of the linearized equation looks as follows [62]:

$$\rho(\vec{k},t) = \rho(\vec{k})\exp(R(\vec{k},t)), \tag{1.12}$$

$$R(\vec{k},t) = -Mk^2\left[\left(\frac{\partial^2 f}{\partial\rho^2}\right)_0 + 2K_\rho k^2\right]. \tag{1.13}$$

It is known that the second derivative of free energy on density is positive everywhere in the area above the spinodal curve [20], and it is negative in the area below the spinodal line. In this area, $R(k)$ will be positive if the wave vector $\vec{k}$ is smaller than the critical value k_c:

$$k < k_c = \sqrt{\frac{-(\partial^2 f/\partial\rho^2)_0}{2K_\rho}}. \tag{1.14}$$

In this case, the density fluctuations with the wavelengths $\lambda > \lambda_c = 2\pi/k_c$ will grow exponentially. The maximum grow rate will be associated with the wave vector:

$$k_m = kc. \tag{1.15}$$

The diffusion Equation 1.11 was tuned in [61] where a hyperbolic diffusion equation was derived:

$$\left(1+\tau\frac{\partial}{\partial t}\right)\frac{\partial\rho(\vec{r},t)}{\partial t} = M\nabla^2\pi\mu, \tag{1.16}$$

where:

τ is the relaxation time within which the system comes to a local equilibrium

μ is the chemical potential of the system

Solution of the hyperbolic Equation 1.16 gives the following result:

$$k_m = \sqrt{(1+k_c)^{1/2} - 1}, \tag{1.17}$$

The last expression yields a better comparison to numerical calculations obtained by MD.

MD consists of numerically solving the classical Newton equations of a physical system of N atoms (molecules) on computers and obtaining the trajectories for the atoms in time. In [61], 1372 atoms interacting via a spherical two-body Lennard-Jones (LJ) potential were studied by MD using a canonical ensemble. According to statistical mechanics, the system temperature T, which was defined as an average kinetic energy per atom, is the conservative variable for such an ensemble.

The cut-off radius of the LJ potential was set at a distance of 2.5σ, where σ is the effective diameter of the atom. Atoms were placed in a cube with sides of $L = 15.8\sigma$, which corresponded to an average atomic density of $\rho\sigma^3 = 0.35$, which was close to the critical density of an LJ fluid. Infinite conditions were modeled by introducing periodic boundary conditions in two of the three-space dimensions. In the beginning of the simulation, a hypothetic LJ fluid was equilibrated at $k_B T/\varepsilon = 2.0$ by placing it in a one-phase region, well above the critical temperature, which is $k_B T/\varepsilon = 1.4$. After equilibration, the system was suddenly cooled down by rescaling the atomic velocities where the velocities were multiplied to a factor equal to the ratio of the current total kinetic energy to that corresponding to a reference temperature of $k_B T/\varepsilon = 0.8$ each fourth time step. Therefore, these quenchings place the atomistic LJ fluid into an area of absolute instability (cf. Figure 1.1). After rescaling the atomic velocities that lasted 2 ps (or 200 time steps), the system started to evolve without any velocity rescaling constraint.

As a result, the initially uniform system quickly transformed into a two-phase system with the characteristic size defined by the length $\lambda = 7.88\sigma$, which is in good agreement with the Kahn-Hilliard theory that predicts $\lambda = 6.7\sigma$ as a wavelength of the oscillation process leading to phase separation after the system quenches.

In [63,64], an MD method was applied to reveal density fluctuations of a two-dimensional LJ fluid experiencing spinodal decomposition. The authors found that the process lasts 30 ps at temperature $k_B T/\varepsilon = 0.45$ and density $\rho\sigma^3 = 0.325$. During this time period, the density fluctuations reach the liquid density and the system decomposes into a two-phase system; during the further evolution, the liquid droplets join into bigger ones. If the earlier stage can be characterized by the dependence of the average cluster size on time $R \sim t^{1/2}$, this changes to a dependence $R \sim t^{1/3}$, at the latest stage, independent of the dimensionality of the system.

1.5   The Method of Molecular Dynamics

The MD method was discovered by Rahman* in 1960 while he was working at Argonne [65]. Rahman calculated the self-diffusion coefficient of liquid argon, and since then MD has gained admiration as a powerful method applicable to many areas of condensed matter for both its thermodynamics and kinetic properties [66–109].

Norman and coauthors have developed a theory of MD that is applicable for areas where classical trajectories of atoms are well defined, as well as in the regions where the motion is semi-classical [66,67].

The MD method has been applied successfully to the detailed study of the structure of liquid and amorphous metals [104,105], the calculation of transport coefficients of dense gases [93], the properties of defects in crystals under irradiation [107], crystallization from a gas phase [100], the calculation of the kinetic coefficients of liquid argon [71,72], studying phase transitions [85,87], and the properties of dense plasma [97,98].

The MD method consists of a numerical solution of a set of the equations of motion of a system of interacting particles in a certain atomistically small (but still macroscopic) system that can exchange heat, energy, or mass with the rest of the world. Several numerical algorithms have been developed where the simplest method is called Verlet's method [77] (Figure 1.2).

In the one-dimensional case, the algorithm is as follows:

$$x_{n+1} = 2x_n - x_{n-1} + h^2 \cdot f(x_n)$$

$$v_n = (x_{n+1} - x_{n-1})/2h, \tag{1.18}$$

where:

x_n, v_n are the position and velocity of a particle at time $t_n = n * h$

h is the time increment

$f(x)$ is the external force acting at the particle at a position x_n

---

* Argonne physicist Annesur Rahman, known worldwide as the "father of molecular dynamics," pioneered the application of computer science to physical systems.

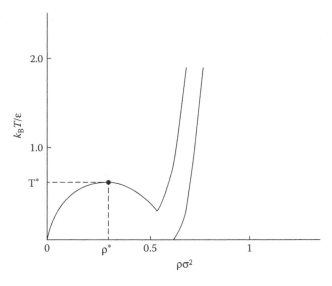

Figure 1.2 Phase diagram of a two-dimensional fluid calculated by a Monte Carlo method in critical values of temperature and density were obtained as follows: $k_B T/\varepsilon = 0.533$, $\rho\sigma^2 = 0.335$.

At the initial time instant, the cluster positions are selected to be at the positions of an ideal poly-facet, if the number of atoms in the cluster is less than 13 [77]. For larger clusters ($n > 50$), it is better to generate the initial structure of the cluster according to a minimum potential energy by using a Monte Carlo procedure where the minimization program starts from a spherical cluster program [77].

In many papers where clusters were used as a projectile bombarding the surface from a gas phase, the clusters were simply cut out spherically from a fragment of FCC lattice, without equilibration [110], however, they can be thermally equilibrated at a given temperature later.

The initial velocities of the cluster atoms were assigned via a Monte Carlo procedure that generates a distribution function of the velocities that is close to the Maxwell function for the reference temperature.

While there is no problem with the temperature definition for large clusters, there is a logical inconsistency of the temperature of small clusters where $n < 10$.

If the cluster is studied in vacuum, the internal temperature can be defined as a kinetic energy per atom of the internal degrees of freedom E_k averaged over a large time interval [78]:

$$k_B T = \frac{2\langle E_k \rangle}{3n - 6},$$
(1.19)

where n is the number of atoms in the cluster.

In [67,68], the authors introduced an averaging over the initial configurations of an arbitrary MD system that had not attained equilibrium and, therefore, the ergodic theorem on the replacement of the averaging over time with averaging over ensemble is just not valid. The initial configurations for averaging should have been selected statistically independent. In such an approach, the kinetic properties will be obtained as average values over samplings.

The choice of various ensembles in MD is closely related to the definition of the temperature and other thermodynamic variables, such as pressure P and entropy S, in the system [77,78].

One of the most popular ensembles is the so-called microcanonical ensemble, where the equations of motion are solved for a conservative system with constant total mechanical energy.

Nose [80] proposed a canonical ensemble for MD where, in addition to real variables such as position and velocity, one introduces a fictitious variable. This method produces a canonical ensemble for the systems in equilibrium, specifically for the study of small systems of clusters [78]. A deficiency of this otherwise successful method is the nonphysicality of the introduction of a fictitious variable that should be assigned a mass and a velocity.

From the viewpoint of simplicity and adherence to physical meaning, the best method in our mind is the stochastic ensemble based on the solution of the Langevin equations of motion [66,111]. In this method, a dissipative and a stochastic term are added into the right side of the equation of motion (EOM), in addition to regular potential terms. For one-dimensional case, the equation of motion has the following form:

$$m\ddot{x} = f - \gamma m \dot{x} + R,$$
(1.20)

where:

m is the mass
γ is the friction coefficient

    f is the potential force
    R is the random force

It is assumed that the fluctuation-dissipation theorem is fulfilled:

$$\langle R(t) \rangle = 0$$

$$\langle R^2(t) \rangle = 2\gamma m k_B T. \tag{1.21}$$

The random force is chosen to be of a Gaussian type.

The boundary conditions in the MD method are usually chosen to adequately represent a physical character of the problem of interest. If, for example, the physical system is an isolated small cluster, the boundaries of such a system can be a sufficiently large sphere that confines the small system and ideally reflects back the atoms colliding with the sphere from the inside [35] or generating a Maxwellian velocity distribution, or simply generating random velocities [77].

Completely different boundary conditions (BCs) should be relevant for the study of atomic particle collisions with solid surfaces. The BCs in this case should provide energy and heat at the system's boundaries separating the system and the ambient media [112–116], and in addition the shape of the BC should comply with the symmetry of the basic cell.

In the MD method, the particles (atoms or molecules) interact with each other via a force that can be calculated by taking the first derivative of the potential energy of two atoms, which usually can be fitted by a model function. One such widely used model function is the Lennard-Jones potential [6–12], which correctly represents the force field between two noble gas atoms, such as argon, krypton, and xenon:

$$U(r) = 4\varepsilon \left[\left(\frac{\sigma}{r} \right)^{12} - \left(\frac{\sigma}{r} \right)^6 \right], \tag{1.22}$$

where:
    r is the distance between the two atoms
    σ is the effective atomic diameter
    ε is the minimum value of the interaction potential well for two atoms

For metallic systems, the preferable functional form of the interaction potential is the embedded atom method (EAM) type that efficiently takes into account N-body interactions, where N is the number of nearest atoms to the coordination sphere:

$$U(r_{ij}) = A \exp\left[-p\left(\frac{r_{ij}}{r_0}-1\right)-\left\{\sum_j B \exp\left[-2q\left(\frac{r_{ij}}{r_0}-1\right)\right]\right\}^{1/2}\right], \quad (1.23)$$

where A, B, r_0, p, and q are the adjustable parameters that are chosen by fitting the function Equation 1.23 to a large set of experimental data such as cohesion energy, elastic constants, sound speed, equation of state, phase coexistence, and defect properties of the material [72,81].

All calculations in MD are conducted in a system of dimensionless "reduced" units (the so-called MD units) that are built based on a few basic parameters such as atomic diameter σ for the length, mass m, and depth of the potential ε for the energy unit. All of the remaining units can be derived from the basic as follows: time unit, $[\tau] = \sigma (m/\varepsilon)^{1/2}$; velocity, $[v] = (\varepsilon/m)^{1/2}$; and pressure, $[P] = \varepsilon/\sigma^3$ [86].

An important factor in the thermodynamics and kinetics of small clusters is the definition of clusters that allows distinguishing clusters of different sizes in a large mixture of different sizes and calculating the distribution function of the clusters over the sizes. Therefore, the criterion of the cluster should be able to clearly separate various transient aggregates in the fluid from those that we call clusters. The easiest way to do that is to count the number of groups of closely located atoms in space—this is the so-called "geometric" criterion. According to this criterion, the atoms in a cluster should be closer to each other than a certain cut-off radius [46–48,52,53,77].

A more elaborated and more reliable criterion is the "time" criterion, according to which a cluster is a group that stays together for a certain time period without going apart. For example, the period of time for this criterion could be one oscillation period of atomic vibrations [57–59]. The second time criterion should be applied in addition to the geometric one. It should be mentioned that both criteria remarkably overestimate the number of clusters. This was noticed by Tully [117], who studied gas atoms sticking to surfaces. He proposed a new "energy" criterion to better describe the physics of binding two atoms into a group: a bond between two atoms is considered stable if the total relative energy of interaction per this degree of freedom is negative or less than several $k_B T$. The threshold energy for the stable bonding can be found by trials and visualizations of the simulation results.

1.6   Summary of Chapter 1

1. Classical Nucleation Theory (CNT) gives fair predictions of the supersaturation level in diffusion chambers and for condensation in supersonic jets in the presence of a carrier gas.

2. CNT in principle is not applicable for describing nucleation in dense media, because it does not take into account the diffusion characteristics.

3. The major contribution to the calculation of the kinetics of cluster formations based on microkinetic equations comes from the growth and evaporation coefficients of clusters. MD can be used to calculate the missing kinetic coefficients.

4. In the case of direct simulation of the kinetics of cluster formation in dense media by the MD method, is missing a reliable criterion of cluster definition that identifies the atoms belonging to clusters is missing.

5. For studying interactions of gas particles with surfaces, the most important tasks should include the development of models of solid surfaces that take into account energy and heat release at the impacts.

6. The atomistic theory of nucleation of a new phase on solid surfaces cannot calculate the number of clusters of various sizes, especially at the initial stage of condensation.

7. The MD method is widely used in nucleation and condensation studies. However, it is insufficient for the direct simulation of the kinetics of new phase formations in various dense systems.

References

1. J. W. Gibbs, *The Scientific Papers of J. Willard Gibbs*, vol. 1, pp. 55–353, Dover: New York, 1961.

2. M. Volmer and A. Weber, Keimbildung in übersättigten Gebilden, *Z Phys Chem (Leipzig)* 119, 277, 1926.

3. M. Volmer, *Kinetik der Phasenbildung*, Verlag Theodor Steinkopff: Dresden, 1939.

4. R. Becker and W. Döring, Kinetische Behandlung der Keimbildung in übersättigten Dämpfen, *Ann Phys (Leipzig)* 24, 719, 1935.

5. Ya. B. Zel'dovich, К теории образования новой фазы. Кавитация, *Zh. Exper Theor Fiz* 12, p. 525, 1942 (in Russian); English translation: in Ya. B. Zel'dovich, To the theory of the formation of a new phase. Cavitation, *Acta Physico-Chim. (URSS)* 18(1), 1943.

6. Ya. I. Frenkel, *Kinetic Theory of Liquids*, Oxford University Press: Oxford, UK, 1946.

7. D. Turnbull and J. C. Fisher, Rate of nucleation in condensed systems, *J Chem Phys*, 17, 71, 1949.

8. D. Hirth and G. Pound, *Evaporation and Condensation*, MacMillan: New York, 1963.

9. A. C. Zettlemoyer, *Nucleation*, Dekker: New York, 1969.

10. B. J. C. Wu, P. P. Wegener, and G. D. Stein, Condensation of sulfur hexafluoride in steady supersonic nozzle flow, *J Chem Phys* 68, 308–318, 1978.

11. J. J. Burton, On the validity of homogeneous nucleation theory, *Acta Metallurgic* 21, 1225–1232, 1973.

12. F. F. Abraham, *Homogeneous Nucleation Theory*, Academic Press: New York, 1974.

13. B. M. Smirnov, R. S. Berry, and S. R. Berry, *Phase Transitions of Simple Systems*, Springer: New York, 2008.

14. A. A. Lushnikov and A. G. Sutugin, Present state of the theory of homogeneous nucleation, *Russ Chem Rev* 45(3), 197–212, 1976.

15. B. F. Gordiets, L. A. Shelepin, and Yu. S. Shmotkin, Kinetics of isothermal processes of homogeneous condensation, *J Sov Laser Res* 7(6), 588–616, 1986.

16. U. G. Pirumov and G. S. Rosliakov, *Gas Flow in Nozzles*, Springer: Berlin, 1986.

17. Ya. I. Tokar' and A. G. Sutugin, Microkinetic description of nonstationary homogeneous nucleation, *Colloid J USSR* 50(1), 162–165, 1988.

18. B. D. Shizgal and David P. Weaver (Eds), *Rarefied Gas Dynamics: Experimental Techniques and Physical Systems*, vol. 158, Chapter 5, Astronautics and Aeronautics (AIAA): Washington, DC, 1994.

19. H. Vehkamäki, *Classical Nucleation Theory in Multicomponent Systems*, Springer: Berlin, 2006.

20. V. P. Skripov, *Metastable Liquids*, Wiley: New York, 1974.

21. V. P. Skripov, E. N. Sinitsyn, P. A. Pavlov, G. V. Ermakov, G. N. Muratov, N. V. Bulanov, and A. G. Baidakov, *Thermophysical Properties of Liquids in the Metastable (Superheated) State*, Gordon: New York, 1988.

22. V. P. Skripov and M. Z. Faizullin, *Crystal-Liquid-Gas Phase Transitions and Thermodynamic Similarity*, Wiley: New York, 2006.

23. P. Jena, B. K. Rao, S. N. Khanna, and D. Reidel (Eds), *Physics and Chemistry of Small Clusters*, NATO ASI Series B Physics, vol. 158, Plenum: New York, 1987.

24. P. Jena, S. N. Khanna, and B. K. Rao (Eds), *Physics and Chemistry of Finite Systems: From Clusters to Crystals*, vols. 1 and 2, Kluwer: Dordrecht, 1992.

25. P. Jena and S. N. Behera, Clusters and nanostructured materials, in *Proceedings of the International Workshop on Clusters and Nanostructured Materials, Puri, India (1996)*, Nova Science: New York, 1996.

26. J. Jellinek (Ed), *Theory of Atomic and Molecular Clusters*, Series: Springer Series in Cluster Physics, 1999.

27. Clusters and Nanomaterials Theory and Experiment Series: Springer Series in Cluster Physics, Y. Kawazoe; T. Kondow; K. Ohno. (Eds), 2002.

28. P. Milani and S. Iannotta, *Cluster Beam Synthesis of Nanostructured Materials*, Springer Series in Cluster Physics, Series Editors: R. S. Berry, A. W. J. Castleman, H. Haberland, J. Jortner, T. Kondow, 1999.

29. M. A. Sharaf and R. A. Dobbins, A comparison of measured nucleation rates with the predictions of several theories of homogeneous nucleation, *J Chem Phys* 77, 68, 1517–1526, 1982.

30. P. E. Wagner and R. Stray, Measurements of homogeneous nucleation rates for nnonane vapor using a twopiston expansion chamber, *J Chem Phys* 85, 6192–6196, 1986.

31. D. Golomb, R. E. Good, and R. F. Brown, Dimers and clusters in free jets of argon and nitric oxide, *J Chem Phys* 52, 1545–1551, 1970.

32. T. W. Milne and F. T. Greene, General behavior and equilibrium dimer concentration, *J Chem Phys* 47, 4095–4101, 1967.

33. S. K. Aizatullin, I. S. Zaslonko, V. N. Smirnov, and A. G. Sutugin, Study of condensation of iron vapors decay in shock waves, *Chem Phys* 4, 851–856, 1985 (in Russian).

34. A. Dillman and G. E. A. Meier, Homogeneous nucleation of supersaturated vapors, *Chem Phys Lett* 160, 71–74, 1989.

35. V. M. Bedanov, V. S. Vaganov, G. V. Gadiyak, and G. G. Kodenev, Numerical simulation of evaporation of Lennard-Jones clusters and calculation of the nucleation rate in supersaturated vapor, *Sov J Chem Phys* 6, 1960–1967, 1987.

36. V. M. Bedanov, V. S. Vaganov, G. V. Gadiyak, G. G. Kodenyov, and E. A. Rubakhin, Experimental determination of the number of atoms in the critical nuclei, *Chem Phys* 7, 555–563, 1988.

37. V. M. Bedanov, Computer simulation of cluster decay, *Mol Phys* 69, 1011–1024, 1990.

38. A. V. Bogdanov, Y. E. Gorbachev, G. V. Dubrovsky, A. L. Itkin, and E. G. Kolesnichenko, Equilibrium solutions of quasichemical condensation model, *Preprint of A Ioffe Physics & Techn Inst of the Acad Sci USSR* 1163, 40, 1987 (in Russian).

39. I. Napari and H. Vehkamaki, The role of dimers in evaporation of small argon clusters, *J Chem Phys* 121, 819–822, 2004.

40. Yu. G. Korobeinikov, Calculation of three-body rate constants, combustion, explosion and shock waves, *Phys Comb Exp (Fizika Goreniya i Vzryva)* 737–743, 1975.

41. A. A. Vostrikov, Yu. S. Kusner, A. K. Rebrov, and B. G. Semyachkin, Formation of dimers in molecular gases, *Lett J Techn Phys* 3(24), 1319–1323 (in Russian).

42. D. J. Frurip and S. H. Bauer, Homogeneous nucleation in metal vapor. 4. Cluster growth rates from light scattering, *J Phys Chem* 81(10), 1007–1015, 1977.

43. S.-N. Yang and T.-M. Lu, The sticking coefficient of Ar on small Ar clusters, *Solid State Commun* 61(6), 351–354, 1987.

44. W. G. Dorfield and J. B. Hudson, Condensation on CO_2 free jet expansions. 2. Growth of small clusters, *J Chem Phys* 59(3), 1261–1265, 1973.

45. A. P. Godfried and I. F. Silvera, Raman studies of Argon dimers in a supersonic expansion, 2. Kinetics of dimer formation, *Phys Rev A* 27, 3019–3030, 1983.

46. J. W. Brady, J. D. Doll, and D. L. Thompson, Cluster dynamics: A classical trajectory study of A+An → A*n+1, *J Chem Phys* 71(6), 2467–2472, 1979.

47. J. W. Brady, J. D. Doll, and D. L. Thompson, Cluster dynamics: Further classical trajectory studies of A+An → A*n+1, *J Chem Phys* 73(6), 2767–2772, 1980.

48. J. W. Brady, J. D. Doll, and D. L. Thompson, Cluster dynamics: A classical study of A*n → An-1 + A, *J Chem Phys* 74(2), 1026–1028, 1981.

49. P. J. Robinson and K. A. Holbrook, *Unimolecular Reactions*, Wiley: London, UK, 1972.

50. S. F. Chekmarev and F. S. Lyu, Some aspects of dynamic chaos in small Lennard-Jones clusters, *Z Phys D* 20, 231–233, 1991.

51. F. S. Lyu and S. F. Chekmarev, Application of the molecular dynamics methods for studying characteristics of small clusters, *Preprint of Inst Therm Phys* 89, 25, 1988.

52. B. M. Bedanov, V. S. Vaganov, and G. V. Gadiyak, *Modeling Cluster Evaporation in Vacuum*, Reprint of IGG RAS, 9, 20, 1985.

53. B. M. Bedanov, V. S. Vaganov, G. V. Gadiyak, G. G. Kodenev, V. V. Kuznetsov, and E. A. Rubakhin, Equilibrium vapor pressure over a nucleus of clusters, *Sov J Chem Phys* 6, 1960–1967, 1987.

54. W. C. Schieve and H. W. Harrison, Molecular dynamics study of dimer formation in three dimensions, *J Chem Phys* 61, 700–703, 1974.

55. M. Yamashita, T. Sato, and S. Kotake, Kinetic analysis of dimer formation of rare gases, *J Chem Phys* 61, 700–703, 1974.

56. A. V. Krestinin, V. N. Smirnov, and I. S. Zaslonko, Modeling of $Fe(CO)_5$ decomposition and iron atom condensation behind shock waves, *Sov J Chem Phys* 8(3), 689–703, 1991.

57. E. E. Polymeropoulos and J. Brickmann, Molecular dynamics study of the formation of argon clusters in the compressed gas, *Chem Phys Lett* 92, 59–63, 1982.

58. E. E. Polymeropoulos, P. Bopp, J. Brickmann, L. Jansen, and R. Block, Molecular-dynamics simulations in systems of rare gases using Axilrod-Teller and exchange three-atom interactions, *Phys Rev A* 31, 3565–3569, 1985.

59. A. A. Belov and Yu. E. Lozovik, Molecular dynamic study of cluster formation in gases, *Preprint of Inst Spectroscopy of RAS* 159, 1–40, 1988.

60. J. W. Cahn, Phase separation by spinodal decomposition in isotropic systems, *J Chem Phys* 42(1), 93–99, 1965.

61. F. F. Abraham, On the thermodynamics, structure and phase stability of the nonuniform state, *Phys Rep* 59, 92–156, 1979.

62. J. W. Cahn and J. E. Hilliard, Free energy of a nonuniform system. I. Interfacial free energy, *J Chem Phys* 28, 258–267, 1958.

63. S. W. Koch, R. C. Desai, and F. F. Abraham, Dynamics of phase separation in two-dimensional fluids: Spinodal decomposition, *Phys Rev A* 27(4), 2153–2167, 1983.

64. S. W. Koch and R. Liebman, Comparison of molecular dynamics and Monte-Carlo computer simulation of spinodal decomposition, *J Stat Phys* 53(1), 31–41, 1983.

65. A. Rahman, Correlation in the motion of atoms in liquid argon, *Phys Rev* 136, A406–A411, 1964.

66. G. E. Norman, V. Yu. Podlipchuk, and A. A. Valuev, Equations of motion and energy conservation in molecular dynamics, *Mol Simul* 9(6), 417–424, 1993.

67. A. A. Valuev, G. E. Norman, and V. Yu. Podlipchuk, *Molecular Dynamics Method: Theory and Applications, Mathematical Modelling*, Nauka: Moscow, 1989.

68. D. Levesque, L. Verlet, and J. Kurkijarvi, *Phys Rev A* 7, 1690, 1973.

69. D. Levesque and W. T. Ashurst, *Phys Rev Lett* 33, 277, 1974.

70. A. N. Lagar'kov and V. M. Sergeev, *High Temp* 11, 460–464, 1973.

71. A. N. Lagar'kov and V. M. Sergeev, *High Temp* 11, 1039–1043, 1973.

72. A. M. Evseev and A. K. Ashurov, *Russ J Phys Chem* 48, 1263, 1974.

73. A. M. Evseev and A. K. Ashurov, *Moscow Univ Chem Bull* 29(4), 27, 1974.

74. W. W. Wood and J. J. Erpenbeck, Molecular dynamics and Monte Carlo calculations in statistical mechanics, *Ann Rev Phys Chem* 27, 319–348, 1976.

75. A. L. Tsykalo and M. M. Kontsov, Effect of trifold non-additive interactions on thermodynamic properties of dense gases and liquids, *Sov Phys Techn Phys* 47(12), 2601–2607, 1977.

76. A. L. Tsykalo and M. M. Kontsov, Dynamics of atomic movement and self-diffusion coefficients of simple liquid with pair-nonadditive particle interaction, *Sov Phys Techn Phys* 49(7), 1558–1561, 1979.

77. M. P. Allen, Algorithms for Brownian dynamics, *Mol Phys* 47(3), 599–601, 1982.

78. J. G. Gay and B. J. Berne, Computer simulation of Coulomb explosions in doubly charged Xe microclusters, *Phys Rev Lett* 49(3), 194–198, 1982.

79. H.-P. Cheng and R. S. Berry, Surface melting of clusters and implications for bulk matter, *Phys Rev A* 45(11), 7969–7980, 1992.

80. S. Nose, A molecular dynamics method for simulation in the canonical, *Mol Phys* 52(2), 255–268, 1984.

81. S. Sawada and S. Sugano, Structural fluctuations and atom-permutation in transition-metal clusters, *Z Phys D* 14, 247–261, 1989.

82. D. W. Heermann, Computer simulation methods in theoretical physics, *Appl Opt* 26(10), 1818, 1987.

83. H. Müller-Krumbhaar, 5. Modeling small systems, In K. Binder (Ed.), *Monte Carlo Methods in Statistical Physics*, Springer: Berlin, 1979.

84. R. D. Etters and J. Kaelberer, Thermodynamic properties of small aggregates of rare-gas atoms, *Phys Rev A* 11(3), 1068–1079, 1975.

85. S. Zheludkov and Z. Insepov, On thermodynamics of small clusters near the coexistence line of gas-liquid, *Thermophys High Temp* 25, 607–609, 1987 (in Russian).

86. S. Zheludkov and Z. Insepov, Molecular dynamic modeling of kinetics of cluster formation in a supersaturated vapor, *J Phys Chem* 61, 1109–1111, 1987 (in Russian).

87. V. Valuev, S. Zheludkov, Z. Insepov, and V. Yu. Podlipchuk, Molecular dynamic modeling of kinetics of condensation of a supersaturated vapor in a buffer gas, *J Phys Chem* 63, 1469–1477, 1989 (in Russian).

88. Z. Insepov and S. Zheludkov, Molecular kinetics of cluster formation in the dense fluids, *Z Phys D* 20, 453–455, 1991.

89. Z. Insepov and E. Karataev, Calculation of the growth and evaporation constants behind the shock wave front by molecular dynamics, *Math Model* 5, 42–44, 1993.

90. Z. Insepov, E. Karataev, and G. E. Norman, The kinetics of condensation behind the shock wave front, *Z Phys* 20, 449–451, 1991.

91. Z. Insepov and E. Karataev, Molecular dynamics simulation of infrequent events: Evaporation from cold metallic clusters, *Phys Chem of Finite System*, NATO ASI Series 1, 423–427, 1992.

92. Z. Insepov and E. Karataev, Molecular dynamics model of evaporation rate from cold clusters, *Techn Phys Lett* 17, 36–40, 1991 (in Russian), English translation: *Sov Techn Phys Lett*.

93. A. M. Bazhenov and V. N. Desyatnik, Determination of the thermal diffusivity in a molecular dynamic experiment: Thermal diffusivity of fused NaCl, *High Temp* 21(4), 528–530, 1984.

94. V. G. Baidakov and A. E. Galashev, Stability of superheated crystal in molecular-dynamics model of argon, *Sov Phys Sol St* 22(9), 2681, 1980.

95. M. P. Allen and D. J. Tildesley, *Computer Simulation of Liquids*, p. 385, Oxford University Press: Oxford, England, 1987.

96. S. P. Protsenko and V. G. Baidakov, Densities of coexisting phases in the micro-heterogeneous liquid gas system, *High Temp* 26(2), 183–187, 1988.

97. F. B. Baimbetov and T. Ramazanov, Dielectric permittivity of and bremsstrahlung in dense plasma, *High Temp* 30(5), 1025–1028, 1992.

98. F. B. Baimbetov and K. T. Nurekenov, Viscosity and heat conductivity of weekly ionized plasma, *High Temp* 30(6), 1217–1220, 1992.

99. D. Frenkel and B. Smit, *Understanding Molecular Simulation: From Algorithms to Applications*, p. 444, Academic Press: New York, 1996.

100. J. M. Haile, *Molecular Dynamics Simulation*, p. 489, Wiley: New York, 1997.

101. M. Griebel, S. Knapek, and G. W. Zumbusch, *Numerical Simulation in Molecular Dynamics*, p. 470, Springer: Berlin, 2007.

102. R. J. Sadus, *Molecular Simulation of Fluids*, p. 523, Elsevier: Amsterdam, 1999.

103. T. Schlick, *Molecular Modeling and Simulation*, p. 723, Springer: Berlin, 2010.

104. V. A. Polukhin, V. F. Ukhov, and M. M. Dzugutov, *Computer Modeling of the Dynamics and Structure of Liquid Metals* (in Russian), Nauka: Moscow, 1981.

105. V. A. Polukhin and N. A. Vatolin, *Modeling of Amorphous Metals* (in Russian), Nauka: Moscow, 1985.

106. A. N. Lagar'kov and V. M. Sergeev, Molecular dynamics method in statistical physics, *Sov Phys Uspekhi* 21, 566–588, 1978.

107. V. V. Kirsanov and A. N. Orlov, Computer simulation of the atomic structure of defects in metals, *Sov Phys Uspekhi* 27(2), 106–133, 1984.

108. R. W. Hockney and J. W. Eastwood, *Computer Simulation Using Particles*, Taylor & Francis: New York, 1988.

109. D. V. Fedoseev, R. K. Chuzhko, and A. G. Grivtsov, *Heterogeneous Crystallization from the Gaseous Phase* (in Russian), Nauka: Moscow, 1978.

110. H. Hsie and R. S. Averback, Molecular dynamics simulations of collisions between energetic clusters of atoms and metal substrates, *Phys Rev B* 45, 4417–4430, 1992.

111. J.-O. Bovin and J.-O. Malm, Atomic resolution electron microscopy of small metal clusters, *Z Phys D* 19, 293–298, 1991.

112. J. C. Tully, Dynamics of gas–surface interactions: 3D generalized Langevin model applied to fcc and bcc surfaces, *J Chem Phys* 73, 1975, 1980.

113. M. Shugard, J. C. Tully, and A. Nitzan, Dynamics of gas–solid interactions: Calculations of energy transfer and sticking, *J Chem Phys* 66, 2534, 1977.

114. G. Q. Xu, R. J. Holland, S. L. Bernasek, and J. C. Tully, Dynamics of cluster scattering from surfaces, *J Chem Phys* 90, 3831, 1989.

115. C. W. Muhlhausen, L. R. Williams, and J. C. Tully, Dynamics of gas–surface interactions: Scattering and desorption of NO from Ag (111) and Pt (111), *J Chem Phys* 83, 2594–2606, 1985.

116. R. R. Lucchese and J. C. Tully, Trajectory studies of vibrational energy transfer in gas–surface collisions, *J Chem Phys* 80, 3451, 1984.

117. E. K. Grimmelmann, J. C. Tully, and E. Helfand, Molecular dynamics of infrequent events: Thermal desorption of xenon from a platinum surface, *J Chem Phys* 74, 5300, 1981.

CHAPTER 2

Molecular Models of Cluster Formation

2.1 Introduction

In Chapter 1, we showed that cluster formation is a fundamental phenomenon that occurs in a wide variety of physical and chemical systems. They include a large variety of physical systems such as rarefied atomic and molecular gases, dense gases, liquids, adsorption layers on solid surfaces produced during molecular beam epitaxy (MBE), and chemical vapor depositions (CVD).

Examples of such systems are adiabatically expanding gas jets through submillimeter nozzles, dense gases near coexistence states with liquids, binary solid solutions in which the concentration of the solute is higher than the solubility threshold, local and extended defects in irradiated solid targets, and phase transitions in the adsorption layers on solid surfaces interacting with gas or plasma by particle/energy exchange.

Since cluster formation is a common precursor (a preliminary step) in so many physical phase transformations, our theoretical study starts with a systematic definition of the physical cluster itself. We will call an atomic aggregate a physical cluster if it is confined in space via attractive nonsaturated (or physical, or pair-additive) forces or if it exists longer than several vibration periods in time. According to these two definitions, it is straightforward to introduce two criteria of a physical cluster: (1) a combination of geometry and energy criteria where a closely confined group of atoms (molecules) in space can be considered a cluster if the total energy of relative motion of the group is negative and is larger than some threshold that depends on temperature; (2) a combination of geometry and time criteria that consider a group a cluster if it resides longer than several vibration periods of the atomic (molecular) group.

In this section, we develop molecular models of cluster formation and evaporation.

2.2   Cluster Identification Criterion

Cluster identification criteria were introduced for the first time by the researchers who later developed a Monte Carlo (MC) method [1,2]. Since the MC method uses the atomic positions only, these criteria were purely geometric: a group of atoms (molecules) was identified as a cluster, if the atoms in the group were closely held together in space. Therefore, the MC criterion could have been expressed as follows:

$$r_{ij} < R_{cut,} \tag{2.1}$$

where:

r_{ij} are the distances between the atoms in the group, $(i, j = 1, 2, ...N)$

R_{cut} is the threshold distance

This criterion overestimated the number of clusters, since some of the closely located atoms were able to have large kinetic energies.

The MD method operates with both coordinates and velocities of all the atoms in the system; therefore the criterion can be extended by including the relative total energy of a geometrically close group of atoms and finding a cluster with the negative total energy that ensures the stability of the cluster in time.

Brinkman and coauthors extended the geometry criterion used in MC by adding a procedure to check the stability of the cluster within one oscillation period [3,4]. This criterion can be called a time one.

A new cluster identification criterion was introduced in [5] that includes the total relative energy of the system of closely located atoms at distances closer than the certain cut-off distance. Therefore, the new criterion was accordingly named an "energy criterion." The following algorithm was used for the energy calculation: at the first stage, the groups of closely located atoms were obtained in accordance with Equation 2.1. These groups are as follows: m_1—monomers (single particles); m_2—dimers of atoms; m_3—trimers of atoms; and so on. A sphere of radius R_c centered at an atom i ($i = 1, 2, 3,..., N$) is drawn and all particles $j \neq i$ inside the sphere are stored into a list of neighbors of the i-atom and each j-particle is given a suitable flag.

After revealing all of the particles included in one of the above groups, the program moves to the particles that are not listed in any group and the procedure is repeated again until all particles are tested against involvement into a group. Such an algorithm is very fast; however, it is accurate only for rarefied gases, where the number of large clusters is small. For dense gases, there is an added procedure that checks the energy of the atoms that are already inside a geometrical group of closely located atoms.

The energy of the groups of atoms is then analyzed:

$$E_{rel} = \frac{m}{2} \sum_{i=1}^{n} v_{rel,i}^2 + \sum_{i<j} U_{ij} < 0, \tag{2.2}$$

where:

v_{rel} is the velocity of an i-th atom relative to the center of mass of n-atomic group

U_{ij} is the potential energy of atoms i and j

The undefined parameter R_c is introduced as follows: after the system attains a certain preliminary equilibrium state, an average cluster size is calculated in the system

$$\bar{k} = \frac{1}{T} \int_0^T \sum_{k=1}^{k_m} \left[\frac{n_k(t)}{M_{cl}(t)} \right] dt = \frac{1}{T} \int_0^T \bar{k}(t) dt, \tag{2.3}$$

where:

T is the total computing time on the equilibrium interval

k_m is the largest size of the groups

$$M_{cl} = \sum n_k = \sum_{k=1}^{k_m} (m_k - l_k),$$

where l_k is the number of collisional complexes with the relative energy $E_{rel} > 0$

The variable was calculated for several values of R_c and the dependence of $\bar{k}(R_c)$ was obtained.

The calculations revealed the dependence had a linear rise of $\bar{k}$ at small R_c and reached a plateau at a certain R_c. The threshold cut-off radius was selected in the middle of the plateau interval.

As the calculations have shown [5], the energy criterion is sufficiently simple and reliable for both rarefied and dense gases. However, it still should be modified if the clusters are to be revealed at the densities typical for liquids.

2.3   Models of Evaporation, Stabilization, and Excitation of Clusters

New models of calculations of the kinetic coefficients were proposed in the previous sections of this chapter. The following coefficients were studied: k_f is the coefficient of formation of a complex that has an excess energy in the vibration and rotation degrees of freedom; k is the rate of the opposite reaction, the evaporation rate of the atoms from a thermally excited cluster. Schematically, such a reaction can be represented as follows:

$$A_1 + A_n \underset{k_d}{\overset{k_f}{\rightleftarrows}} A_{n+1}^*, \tag{2.4}$$

where:

    A_1 refers to a monomer (single atom)

    A_n is the n-cluster built of n atoms

    the asterisk (*) corresponds to an excited complex

A collision of a monomer with an excited cluster can stabilize the latter, if the monomer takes the excess kinetic energy and lowers the total relative energy of the cluster. The appropriate kinetic coefficient is denoted k_s and the rate of the opposite process is k_e. Obviously, k_e corresponds to an excitation rate in the case where the monomer is not sticking to the cluster. These two processes can be represented as follows:

$$A_1 + A_{n+1}^* \underset{k_e}{\overset{k_s}{\rightleftarrows}} A_1 + A_{n+1}. \tag{2.5}$$

A dimer A_2 is formed by a separate three-particle collision process, with a rate of k_{tr}, with the rate of an opposite reaction of k_{sd} that defines the rate of the impact dissociation of the dimer by a monomer collision:

$$A_1 + A_1 + A_1 + \xrightarrow[k_{sd}]{k_{three}} A_1 + A_2. \tag{2.6}$$

The rate constants are represented as $k_i = \alpha_i S_r$, where α_i are the probabilities of the above reactions and $S_r = \pi p_m^2 \cdot v_T$ is the appropriate kinetic cross sections of collisions; p_m is the largest impact parameter above which there is no collision; and v_T is the relative velocity of collision. The molecular dynamics (MD) method can be used to easily calculate the probabilities $\alpha_f, \alpha_s, \alpha_e,$ and α_{three}.

Figure 2.1 shows the geometry of a collision in MD calculations of the probabilities of the various channels. A cluster containing n atoms is placed so that its center of mass is at the origin of a Cartesian coordinate system and a Maxwellian distribution of velocities, with the temperature T is assigned to the velocities of the atoms in the cluster.

The cluster is bombarded by a single atom (the probe atom) moving with a velocity $v_T = (3k_B T/\mu)^{1/2}$ parallel to the x-axis, at a distance ρ from the x-axis. This distance, called the impact parameter, is an important variable for energy exchange in a collision event.

The methodology of calculation is published in [6,7]. The initial coordinates of the clusters were selected to be equilibrated at a low temperature and were obtained from [8]. The rotation motion of the clusters was eliminated by calculating the energy of rotation degrees of freedom each time step and reducing the atomic velocities to the appropriate amount.

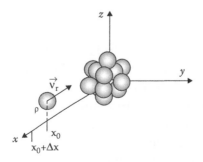

Figure 2.1   Collision model of a monomer (single atom) with the cluster.

The cluster temperature was defined by calculating the total kinetic energy of the relative motion of the atoms relative to the center of mass (cm) of the cluster:

$$
k_B T = \frac{\left\langle \sum_{i=1}^{n} m v_i^2 \right\rangle}{(3n-6)}.
$$

The orientation of the cluster in space was selected randomly according to a uniform probability density. To do that, before each sampling the cluster coordinates were transformed to three random angles relative to the Cartesian coordinate system [8].

The rotation matrixes were selected as follows:

$$
\begin{pmatrix} x' \\ y' \\ z' \end{pmatrix} = \begin{pmatrix} 1 & 0 & 0 \\ 0 & \cos 2\pi\xi_1 & \sin 2\pi\xi_1 \\ 0 & -\sin 2\pi\xi_1 & \cos 2\pi\xi_1 \end{pmatrix} \cdot \begin{pmatrix} x \\ y \\ z \end{pmatrix},
\tag{2.7}
$$

for rotation about the x-axis;

$$
\begin{pmatrix} x'' \\ y'' \\ z'' \end{pmatrix} = \begin{pmatrix} \cos 2\pi\xi_2 & 0 & \sin 2\pi\xi_2 \\ 0 & 1 & 0 \\ -\sin 2\pi\xi_2 & 0 & \cos 2\pi\xi_2 \end{pmatrix} \cdot \begin{pmatrix} x \\ y \\ z \end{pmatrix},
\tag{2.8}
$$

for rotation about the y-axis; and

$$
\begin{pmatrix} x''' \\ y''' \\ z''' \end{pmatrix} = \begin{pmatrix} \cos 2\pi\xi_3 & \sin 2\pi\xi_3 & 0 \\ -\sin 2\pi\xi_3 & \cos 2\pi\xi_3 & 0 \\ 0 & 0 & 1 \end{pmatrix} \cdot \begin{pmatrix} x \\ y \\ z \end{pmatrix},
\tag{2.9}
$$

for rotation about the z-axis. In Equation 2.7 through 2.9, ξ_1, ξ_2, ξ_3 are uniformly distributed in the interval [0,1] of random numbers.

The initial position of the probe atom is selected as follows: the y- and z-positions were uniformly distributed within an area of $2p_m \times 2p_m$ and acceptable $\{y, z\}$ pairs were those restricted by the maximum impact parameter $p_m^2 < y^2 + z^2$. An x-position of the probe atoms was uniformly in the interval $[x_0, x_0 + \Delta x]$ for averaging the results over the phase of collisions. The averaging was realized by the following sampling:

$$x_i = x_0 + \Delta x \xi_4,$$
$$y_i = p_m \xi_5,$$
$$z_i = p_m \xi_6,$$
$$y_i^2 + z_i^2 < p_m^2,$$

(2.10)

where ξ_4, ξ_5, ξ_6 are the uniformly distributed random variables over the interval [0,1].

The MD method calculates the trajectories of all the atoms in the system by solving the Newtonian equations of motion, and therefore it is capable of tracking the outcomes of all the collisions between single atoms with a cluster.

The following rate constants are needed: the excitation probability – α_e; the probability of formation of a larger cluster by association with a single atom – α_f; the stabilization probability – α_s; and impact dissociation – α_{sd}, which can be calculated as appropriate frequencies of the outcomes:

$$\alpha_i = \frac{N_i}{\sum\limits_{j=1}^{M} N_{ij}}, \qquad i = e,e,s,sd,$$

(2.11)

where:

    N_i is the number of events where the outcomes were of i-type

    M is the total number of samples

This model of the reaction rate constants is applicable for calculating chemical reactions, for example, occurring during the formation of iron dimer Fe_2:

$$Fe + FeCO \xrightarrow{\ k_2\ } Fe_2 + CO,$$

(2.12)

where the constant k_2 can also be expressed via a gas-kinetic cross-section Sr and probability α_2:

$$k_2 = \alpha_2 * S_r.$$

2.4   Model of Electronic Excitation of Cluster Atoms

Electronic excitation of cluster atoms should be taken into account at impacts of cluster ions with surfaces with energies above

10 eV/atom [9]. This section presents a molecular model of excitation that is applicable to MD calculations of processes of rare-gas atoms and clusters with surfaces of solid targets [10].

During the collision with a solid surface, a cluster experiences a significant pressure and a very high temperature that are built up inside the cluster. Simple estimates give collision area temperatures and pressures of the order of $T \sim 10^5 \, \text{K}$ and $P \sim 10^7 \, \text{bar}$ [11].

MD was used as a method of calculations in several papers that studied the electronic excitation of the molecules comprising the cluster or the surface [12,13]. In these papers, the process of electronic excitation was modeled by instant switching of the potential energy function to a new one. In our paper [10], the excitation process was modeled in more realistic terms as follows: the initial non-excited molecules were interacting via a more realistic potential of the Buckingham type:

$$U(r) = \varepsilon \left\{ \frac{6}{\alpha - 6} \exp\left[\alpha\left(1 - \frac{r}{r_0}\right) \right] - \frac{\alpha}{\alpha - 6}\left(\frac{r_0}{r}\right)^6 \right\},$$
(2.13)

which reasonably describes the properties of many compressed rare-gas atoms [14].

The excited atoms were interacting with the same potential except that the parameter r_0 was increased by 10 % and ε and α were adjusted to shock wave compression experiments of liquid Argon [14].

The parameters of the interaction potential between the non-excited and excited molecules were determined via the Lorentz–Berthelot mixing rules that estimate the interatomic potential parameters ε_0, r_0, and α for a mixed pair of molecules (ij, where $i \neq j$) by combining the corresponding potential parameters for the two pairs of identical molecules (ii and jj):

$$\varepsilon_{0ij} = \sqrt{\varepsilon_{0ii}\varepsilon_{0jj}},$$

$$r_{0ij} = \frac{r_{0ii} + r_{0jj}}{2},$$
(2.14)

$$\alpha_{ij} = \sqrt{\alpha_{ii}\alpha_{jj}}.$$

The main difference between our method and the one used in the papers [12,13] is that we switched between two potentials according to a Monte Carlo procedure with the probability density

$$P_e = 1 - \exp\left[-\left(k_B T_{cl}/E_e\right)^2\right] \cdot \exp\left[-\beta\left(\vec{r}_i - \vec{r}_{c.m.}\right)\right], \tag{2.15}$$

where:

T_{cl} is the cluster temperature

E_e is a critical excitation energy that was adjusted to the experiment

β is a geometry parameter that controls excitation of the central atoms

$\vec{r}_i$ and $\vec{r}_{c.m.}$ are the radius-vectors of atom i and the center-of-mass of the cluster

Excited atoms return back to the nonexcited state within a certain time delay τ, which is the lifetime of the excited state.

Switching between two interaction potentials was preceded at each time step by the inverse function method proposed by John von Neumann for generating random values for such random quantities as free path and angular distributions [15]. If an atom was found to be in an excited state, the time spent from the beginning of excitation by the atom in this state was calculated and then compared to a random variable distributed in accordance with a radiation decay function. If it was obtained that the lifetime of the excited state was larger than that of a radioactive decay function, then the potential function was switched. A sample code with this algorithm is given in Appendix 2.

2.5 Quasi-Chemical Model of Condensation

The MD model of rate constant calculation proposed in Section 2.2 can be used for determining the distribution function of clusters over sizes and to calculate the nucleation rate of a vapor–liquid phase transition—the characteristic that can be measured in experiment—directly from the interatomic potential, thus without using any adjustable parameter.

To begin this important task, the evolution of cluster sizes can be represented as a set of chemical reactions Equation 2.4 that can be transformed into a set of kinetic rate equations for the concentrations of various clusters.

In this section, a simple kinetic model called the "quasi-chemical model" (QCM) will be introduced which is a simple but versatile model applicable to more complicated processes such as nonisothermal condensation (see, e.g., [16,17]).

According to the QCM, a supersaturated vapor can be represented as a collection of stable monomers A_1 and clusters of various sizes A_n ($n=2...N$) and of excited cluster (transition) states A_n^* ($n=2...N$). A detailed balance between the clusters of various sizes is implemented via bimolecular reactions given by Equations 2.4 through 2.6.

The main assumption of this theory is the quasi-steady state for the concentrations f_n^* of the excited clusters for a certain "critical" size n^* that can be expressed as

$$\frac{df_n^*}{dt} = 0. \tag{2.16}$$

Then the concentrations of the stable clusters can be defined by the following set of kinetic equations:

$$\frac{df_n}{dt} = J_n - J_{n+1}, \tag{2.17}$$

where:

$$J_{n+1} = c_n f_n f_1 - e_{n+1} f_{n+1} \tag{2.18}$$

describes the flux in the space of the size variable. Using the statistical theory of chemical reactions developed by Rice-Ramsperger-Kassel (RRK), the growth rate constants c_n and the decay rate constants e_{n+1} can be related to each other [16]:

$$c_n = c_n^\infty \cdot I_k (n, f_1),$$
$$\tag{2.19}$$
$$e_{n+1} = e_{n+1}^\infty \cdot I_k (n, f_1),$$

where c_n^∞ and e_{n+1}^∞ are the high-density limits of the rate constants which can be expressed as follows [16,18]:

$$c_n^\infty = \alpha_f \cdot S_r,$$
$$\tag{2.20}$$
$$e_{n+1}^\infty = \nu \cdot \exp\left(\frac{-E_0}{k_B T}\right).$$

The Kassel integrals in Equation 2.19 for the high-density limits can be calculated according to [18]

$$I_k\left(n, f_1\right) = \frac{1/(s-1)! \int_0^\infty y^{s-1} e^{-y} dy}{\left[1 + v/k_s f_1 \left(y/y + B\right)^{s-1}\right]},$$

(2.21)

where:

B $= E_0/k_B T$

y $= (E_{n+1} - E_0)/k_B T$

E_{n+1} is the internal energy of a cluster containing $n+1$ atoms

s is the efficient number of oscillators

The quasi-chemical model described above transforms into the Classical Nucleation Theory if one assumes $f_n^* = 0$. The probability of cluster formation in this case will be replaced by a sticking coefficient of an atom and a cluster, if one assumes:

$$\alpha_n = \alpha_f \cdot I_k\left(n, f_1\right).$$

(2.22)

Such an evaluation allows direct comparison of the quasi-chemical model with CNT [19].

2.6   Molecular Dynamics Model of Infrequent Events

Thermal evaporation of atoms from surfaces having characteristic times of the order of microsecond and larger still remain a fast process in experiments. However, this time of thermal desorption or evaporation is much larger than a period of atomic vibration in a crystal lattice that is of the order of 10^{-14} s. The deterministic trajectory simulation methods that are traditionally used in studies of reaction kinetics are based on integrating the classical Newton equations of motion of atoms, and, therefore, the time increment for integration should normally be much smaller than the period of fastest vibrations. Therefore, the desorption processes can be considered as "slow" relative to the fastest atomic vibrations. As a result, the desorbing surface atoms should undertake an enormously large number of attempts before one of the escape trials succeeds and the atom is thermally evaporated (detached) from the surface.

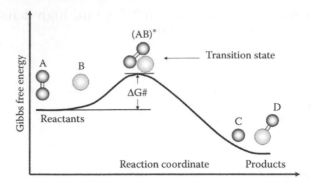

Figure 2.2   Schematic representation of the transition state theory processes. (a) Low Gibbs energy reactant valley contains the reactants A and B that oscillate multiple times without forming bonds between the atoms. (b) A transition complex (AB)* is formed in an infrequent event where the Gibbs activation energy, $\Delta G^{\ddagger}$, is higher than the energy of the reactants. This activation energy defines the potential barrier of the reaction and determines the rate constant of the reaction. (c) The products of the reaction are molecules C and D, with potential energies less than those of the reactants and therefore having larger kinetic energies.

Such processes can conveniently be depicted as a combination of high-frequency oscillations of the atoms in a reactant potential valley (see Figure 2.2) and an infrequent hop of an atom from the reactant valley into the potential valley of the reaction products.

A direct MD calculation of the cluster evaporation rate in the temperature interval of $k_B T/\varepsilon = 0.3$–0.7 was carried out in [20]. This simulation demonstrated that the direct MD simulations of slow thermal evaporation processes at low temperatures become much less efficient and too time consuming. Evaporation of a single atom from a cluster is a similar slow process.

In this section a new model will be developed that combines an MD simulation method with transition state theory (TST) and applied to calculating the evaporation rate for a cold metal cluster [21,22].

The main idea of this new method consists of replacing the actual interaction between the reaction components by a simple analytical function that is easy to integrate, with a compensating potential that was originally developed for calculating the desorption of an atom from a solid surface [23].

It consists of replacing the strong attractive interaction potential of the atoms to the surface V by a sum of two potentials, a strong potential U, which can be fitted by a simple analytical function, and a weak function that depends on the exact positions of the near-surface atoms [23]:

$$W = V - U \qquad (2.23)$$

Since the main task of this chapter is calculating the evaporation rate of a cold cluster, the part U of the total potential V is selected as a central-symmetric function in a continuum media approximation. In the pair-additive approximation, interaction between the two atoms as a Lennard-Jones (12-6) function and the potential U is to be obtained via integration over the volume of the sphere in which the atoms are smeared and the density n_0 of the sphere with radius R is given as the number of atoms divided by the spherical volume (Figure 2.3). The results will be expressed as follows:

$$U(s) = 4\pi n_0 \left[\frac{A}{(s-R)^9} - \frac{B}{(s-R)^3} \right], \qquad (2.24)$$

where:

$A = (1-1/c^9)/45 - (1-1/c^8)/40$

$B = (1-1/c^3)/6 - (1-1/c^2)/4D$, where $c = (s+R)/(s-R)$, $D = s/(s-R)$

The continuum media potential Equation 2.24 will be switching to a known Mie potential (9.3) that describes the interaction between an atom and a solid surface:

$$U_{\text{Mie}}(z) = 27\varepsilon \left[\frac{1}{90} \left(\frac{\sigma}{z} \right)^9 - \frac{1}{12} \left(\frac{\sigma}{z} \right) \right], \qquad (2.25)$$

where z is the distance between the atom and the surface. If R approaches σ, the potential Equation 2.24 transforms into a function very close to a Lennard-Jones type potential.

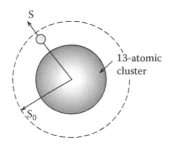

Figure 2.3 Molecular dynamic model of evaporation rate calculation of a 13-atomic metal cluster at low temperatures.

The contribution to the evaporation rate constant that originated from U can be calculated analytically, as will be shown.

The contribution to the rate constant that comes from the residual between the total potential and the analytical part U can be described by a "shallow" potential function W that can be obtained using MD.

The rate constant of cluster evaporation is represented as

$$k_d = k_{TST}F, \qquad (2.26)$$

where

k_{TST} is an equilibrium rate constant of evaporation calculated in the transition state theory

F is the dynamic factor that takes nonequilibrium properties of the evaporation process into account

The dynamic factor can be calculated by MD [18,23].

Let us consider a canonical ensemble of the statistical thermodynamics where the reactants include an n-atomic spherical cluster and a single atom sitting on the cluster surface [23]. The reaction products will be represented by the n-atomic cluster and a single atom separated from the cluster due to evaporation.

Following [23], we introduce a critical surface S as a surface at a distance s from the center of the sphere that divides the phase spaces of the reactants and reaction products.

The flux in the phase space determines the upper limit for the reaction rate. The actual rate constant will be much less than this limit because some of the trajectories of the atom evaporating from the surface will cross multiple times the critical surface and a significant part of them will be turned back at the critical surface S_0.

The reaction rate will be determined by an expression developed in [47]:

$$k_d = \frac{\int d\vec{p}d\vec{q} \int\limits_{0}^{\infty} dv_s P(\vec{p},\vec{q},s_0,v_s)v_s\xi(\vec{p},\vec{q},v_s)}{\int d\vec{p}d\vec{q} \int\limits_{-\infty}^{\infty} dv_s \int\limits_{-\infty}^{s_0} dsP(\vec{p},\vec{q},s_0,v_s)}, \qquad (2.27)$$

where s is an absolute value of the distance between the atom and the cluster surface: $s = |r_0 - R_{cm}|$; this coordinate is normal to the dividing surface at S_0. In our case, the coordinate s is identical to the radial distance from the center of the sphere: $q = r_i - r_j$ $(i, j = 1, \ldots n)$, p – the moments of the atoms, the parameter ξ takes the returning trajectories into account.

$P(p, q, s_0, v_s)$ is the equilibrium probability density in phase space that can be determined by a Hamiltonian function H:

$$P(\vec{p}, \vec{q}, s, v_s) = N \exp\left[\frac{-H(\vec{p}, \vec{q}, s, v_s)}{k_B T}\right],\tag{2.28}$$

and where N is the normalization constant, k_B is the Boltzmann constant, and T is temperature.

The equilibrium part of the rate constant of evaporation k_{TST} can be calculated using the compensation potential method that was developed for calculating the desorption rate of atoms from the solid surface [23].

Equation 2.27 can be written as a product of the two expressions k_{TST} and F as follows:

$$k_{TST} = \frac{\int d\vec{p} d\vec{q} \int_0^\infty dv_s P(\vec{p}, \vec{q}, s_0, v_s) v_s}{\int d\vec{p} d\vec{q} \int_{-\infty}^\infty dv_s \int_{-\infty}^{s_0} ds P(\vec{p}, \vec{q}, s_0, v_s)},\tag{2.29}$$

and

$$F = \frac{\int d\vec{p} d\vec{q} \int_0^\infty dv_s P(\vec{p}, \vec{q}, s_0, v_s) v_s \xi(\vec{p}, \vec{q}, v_s)}{\int d\vec{p} d\vec{q} \int_0^\infty dv_s P(\vec{p}, \vec{q}, s_0, v_s) v_s}.\tag{2.30}$$

The total Hamiltonian can be represented as the sum of two components—the kinetic energy $T(p, v_s)$ and the potential function part $V(q, s)$ —and Equation 2.27 can be integrated as follows:

$$k_{TST} = L s \left(\frac{2 k_B T}{\pi \mu}\right)^{1/2} \frac{g(s_0) \exp\left[-V(\vec{q}, s_0)/k_B T\right]}{\int_{-\infty}^\infty ds\, g(s) \exp\left[-U(s)/k_B T\right]},\tag{2.31}$$

where

$$g(s) = \int d\vec{q} \exp[-W(\vec{q}, s) / k_B T],$$

and $g(s)$ has a meaning of the time spent by the atom at the distance interval from S to $S + dS$ in the potential W.

The parameter $g(s)$ can be calculated by MD.

2.7 Computational Models of Solid Surfaces

In an MD simulation, the equations of motion of interacting particles are solved numerically. Appropriate initial and boundary conditions are supplied. Due to limited computing resources, there is always a problem of choosing a correct size for the model system.

The computational models used for modeling the interactions of atoms and clusters with the surfaces of solid substrates and for studying the kinetics of atomic processes in an adsorption monolayers can be separated into three groups [10,24–29].

The first group of models comprises models in which the substrate atoms are placed at equilibrium positions of an ideal lattice that corresponds to a certain given surface temperature. The excess energy released at the surface by the projectile particle should be removed from the simulation system; this can be done by adding a special friction term in the equation of motion (EOM) of the atoms "for the whole" system. Such EOMs containing friction terms and a random term for energy conservation are called Brownian dynamics [1,2]. Such models were widely used for studies of the interaction of accelerated atoms and clusters with surfaces as well as the kinetics within the absorption layers for studying the cluster formation dynamics in the absorbed layers.

The second group of surfaces models are those that consider the substrate atoms to be located and vibrated each at their equilibrium lattice positions and are absolutely independent from each other. Therefore, only those interactions are taken into account that represent attraction and repulsion between a projectile atom/cluster with individual surface atoms. This model is called the model of independent oscillators [30]. Thermal energy release in such systems is done via adding Langevin forces into the EOM of the top layer. Obviously, this second class of methods is more realistic than that of the first class of models.

The third group of models consists of multizone models that were used for modeling energetic atoms, ions, and clusters with surfaces [24–29].

The primary (central or collisional) zone includes those atoms partitioned into a domain that directly interacts with energetic projectile particles. Therefore, the force and energy changes for these atoms are severe and must be calculated most accurately. MD has been employed to calculate the accurate trajectories for this zone [24–29].

The number of mobile atoms chosen will set up the MD model size: too few atoms may show unphysical effects but increasing the number has a steep price in computation time and required memory as both grow rapidly with an increasing number of degrees of freedom of the system. The usual procedure is to estimate a necessary number of target atoms, often influenced by the available computing power, and to account for the rest of the system by proper boundaries which would allow the flow of energy deposited by the impact to the bulk of the solid.

The size and the shape of this central interaction zone is also determined by the character of the problem under study. For example, the size of this zone can be obtained from the condition that the elastic waves generated at the energetic impact would not be reflected back into the system and hence it would spoil the overall dynamics (Figure 2.4).

One of the often-used techniques, based on stochastic Langevin dynamics, employs two or three atomic layers surrounding the central volume treated by MD. These layers of damped atoms represent a "thermal bath" that simulates heat transfer to the bulk of the target. This technique seems to be reasonable in simulating low-energy processes, such as film deposition. This zone is surrounded by a secondary zone that mimics the thermostat and where the EOM contains Langevin forces and random forces, in addition to interaction forces and energies.

The third zone contains surface atoms fixed at equilibrium positions. They are necessary to stabilize the structure of two internal zones at strong impact of the projectile with the surface. In [29], where the collisions of an energetic cluster ion (with energy of <100 eV/atom) were studied, the instantaneous positions of atoms

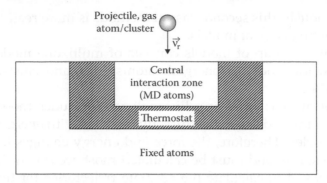

Figure 2.4 Multizone model of surface. (From Z. Insepov and A. Zhankadamova, *Z Phys D* 20, 145–146, 1991; H. Haberland, et al., *Mater Sci Eng B* 19, 31–36, 1993; H. Haberland, et al., *Z Phys D* 26, 229–231, 1993; H. Haberland, et al., *Nucl Instrum Methods Phys Res B* 89, 419–425, 1993; M. Ishii, et al., Molecular dynamics simulation for cluster bombardment of solid surface, in *Proceedings of the 40th Spring Meet, Japan Soc Appl Phys* 2, 588, 1993; Z. Insepov, et al., Molecular dynamics simulation for sputtering of gold surface by an energetic argon cluster, in *Proceedings of the 40th Spring Meet, Japan Soc Appl Phys* 2, 588, 1993.)

in the third zone were chosen to obey the continuum mechanics equations and were capable of responding to the shock waves generated at the impact.

Energetic cluster impacts create violent collisions between atoms in the central zone where equivalent temperature and pressure may reach 10^5 K and 10^6 bars, respectively [31]. For modeling such events, the boundary can be made "flexible" by allowing its expansion to keep the average pressure constant for a given modulus of elasticity. This technique is not completely satisfactory as it uses the average pressure, which depends on the system size, and requires knowledge of materials characteristics, such as thermal conductivity and compressibility. These parameters cannot be reliably extrapolated from the normal equilibrium conditions to the extreme state of matter in the collision zone.

The problem of the boundary conditions also can be examined by considering shock waves created by the energetic cluster impact. Unphysical reflections of the shock waves from the system boundary may show up in MD results, distorting the picture of the investigated process. Shock wave reflections have been revealed for systems as large as 4 3105 target atoms studied by MD if fixed periodic boundary conditions are used, for cluster impact energy as low as 10 kV [32,33]. A new hybrid model utilizing MD for the atoms in the central collision zone, and continuum mechanics and thermodynamics outside, has been proposed in a previous paper and used for a crater formation study in [34,35]. In this model, a shock wave theory is used to establish the minimum size of the central zone, that is, the location of the boundary between the volume where MD models atomic collisions and the outside, which is treated as continuum. This hybrid model successfully solves the problem related to a finite system size by reducing the number of required degrees of freedom, and will be applied to the modeling of craters in the present article. For convenience, some equations of the hybrid model are given next.

According to the shock wave theory [36], the distance L from the center of the cluster impact to where the shock wave becomes thermally equilibrated can be estimated as $L \sim d \times 10^{1/3}$, where d is the cluster diameter. For argon clusters with 100 atoms, $d \approx 30$ Å and L is of the order of 100 Å. The shock wave travels distance L in time $\tau_0 = L/v_0$, where v_0 is the cluster velocity (prior to impact, i.e., an upper bound). For argon clusters with kinetic energy of 20 eV/atom, $v_0 \approx 10^4$ m/s, and the time interval is $\tau_0 \sim 1$ ps. The hybrid model establishes the boundary of MD volume at L and follows the system evolution with MD calculations for the time at least several times τ_0. The volume beyond radius L from the point of impact is treated as a continuum, using a finite-element method to solve the continuum mechanics equations

$$\rho \frac{d^2 u_i}{dt^2} = \frac{\partial \sigma_{ik}}{\partial x_k},$$ (2.32a)

where:

u_j is the displacement vector of the j-th cell
σ_{ik} is the stress tensor
ρ is the solid density [37]

Heat transfer is described by the following linear thermodynamic equation:

$$\frac{dT(r,t)}{dt} = \chi \Delta T(r,t),$$ (2.32b)

with temperature (T) at position r and time t. Equations 2.32a and 2.32b are coupled by the stress tensor dependence on material constants and temperature:

$$\sigma_{ik} = -K\alpha T\delta_{ik} + Ku_{ll}\delta_{ik} + 2\mu\left(u_{ik} - u_{ll}\frac{\delta_{ik}}{3}\right) + \xi\dot{u}_{ll}\delta_{ik} + 2\eta\left(\dot{u}_{ik} - \dot{u}_{ll}\frac{\delta_{ik}}{3}\right)$$

$$u_{ik} = \frac{1}{2}\left(\frac{\partial u_i}{\partial x_k} + \frac{\partial u_k}{\partial x_i}\right), \qquad\qquad (2.32c)$$

where:
- α is the thermal expansion coefficient
- K and ξ the bulk elastic modulus and viscosity
- μ and η are the sheer modulus and viscosity
- u_{ik} is the strain tensor

For silicon at room temperature, $\chi \sim 1$ cm^2/s [38], so that the characteristic heat transfer time $\tau_1 = L^2/\chi = 1$ ps is similar to the characteristic deformation time τ_0. This is convenient for numerical computations as it allows the use of common time steps for calculating thermal and mechanical variables. Equations 2.32a through 2.32c have been used to numerically solve a two-dimensional problem projected by cylindrical geometry defined by the symmetry of the system.

In the MD calculations, we used the Buckingham potential to represent two-body forces between cluster atoms and between the cluster and target atoms while interactions between Si atoms were represented by the Stillinger–Weber potential. The cluster was modeled by cutting a spherical volume from a face-centered cubic argon lattice with initial temperature set to zero.

The clusters used in simulations contained about 100–200 atoms and had kinetic energy of a few tens of kV. The cylindrical target model contained $\sim 10^5$ atoms in the central MD zone, while the continuum mechanics calculations extended to ten times larger volume. In fact, there is no limitation for the system size in this method at all.

A similar goal was had in a model where the impact energy was absorbed by a piston that was capable of absorbing a significant part of the energy on an accelerated ion impact.

2.8 Shock Wave Generation Model
at Energetic Cluster Impact

To investigate the phenomena of shock wave generation, two-dimensional (2D) MD models of energetic gas cluster impacts on solid surfaces have been developed. In our previous paper we revealed that three-dimensional (3D) effects play only a little role in comparison with results of 2D simulation in the case of thin metal film deposition at low and intermediate energies [12]. The same opinion was expressed in [8], where the emission processes occurring at hypervelocity dust particle impacts on surfaces were studied in the frame of 2D hydrodynamics. To apply the results of 2D-modeling to a real 3D situation, the distances have to be rescaled: $x_{3D} = (x_{2D})^{2/3}$. But the pressure and the velocities cannot be found so easily, and the results have a qualitative meaning only.

The initial atomic positions of a 2D cluster and a target were arranged for a hexagonal-close-packed (hcp) structure. The mutual atomic interactions were modeled via Buckingham's two-body potential with equilibrium distances between the atoms of 3.4 and 2.2951 Å, and the energy depths of 10.254 meV and 2.817 eV for Ar and Si atoms, respectively. The 2D cluster sizes were varied between 200 and 350 Ar atoms, which correspond to 3D cluster sizes of 1000–3000 atoms. The instantaneous cluster temperature was calculated from the kinetic energy per atom belonging to the transversal degrees of freedom of the cluster. The initial cluster temperature was set to zero. The initial translational cluster energy varied between 17 and 85 eV per cluster atom. The target sizes were varied in the range of order of 10,000–50,000 atoms. The substrate temperature was chosen to be room temperature by use of the Langevin dynamics (LD) technique.

The expandable boundary conditions of the substrate were used as described in our earlier papers [14].

Average temperatures in front of the shock wave and behind it were calculated from a transversal projection of the velocity of the atoms inside a semi-spherical layer having its center at the impact point, dependent on space and time variables. The local target temperature and pressure were calculated for computational cells containing 25–50 target atoms in accordance with nonequilibrium thermodynamics as in [12]. The shock wave front was assumed to be identical to a compression front obtained from the radial mass-velocities of the target atoms.

The cluster impact induced damage of the target material was estimated via atomic displacements. The atoms, whose displacements were larger than half the lattice constant, had been considered as disordered.

For the late stage of impact, the hydrodynamic variables comply with some general relations that are known as the self-similar behavior of shock waves [39]:

$$R_c \propto t^{\alpha},$$

$$M(R) \propto \rho_0 R \propto t^{\alpha},$$

$$P_{sw} \propto M^{-n} \propto t^{-\alpha n},$$

$$u = \frac{\partial R}{\partial t} \propto t^{\alpha-1} \propto P_{sw}^{1/2} \propto M^{-n/2} \propto t^{-\alpha n/2}$$

$$n = \frac{2(1-\alpha)}{\alpha},$$

where:
  R is the shock wave radius at the time t
  P_{sw} is the pressure behind the shock front
  ρ_0 is the undisturbed target density
  M is the total target mass involved into movement
  u is the mass velocity behind the front

2.9   Sticking Coefficients of the Gas Atoms to Surface

The sticking coefficient of atoms, molecules, and clusters to surfaces was calculated based on the models that were described in Sections 2.2 through 2.7. A projectile atom was placed above the central collision zone at a certain distance z_0. The vertical component of velocity of these atoms can be calculated from the gas temperature as follows:

$$v_z = \sqrt{\frac{k_B T}{m}},$$

where:
  T is temperature
  m is the atomic mass

During each collision of the projectile with the surface, the following outcomes were calculated: (1) the atoms stick, (2) the atoms are reflected, (3) the atoms are considered to be stuck to the surface but have large kinetic energy and therefore could have a probability of being reflected eventually; this class is called a transient (between stuck and reflected).

The following criteria were used for calculations:

1. If the projectile changes the sign of the normal velocity component twice in the course of collision, it is considered as stuck. This criterion seems to be reasonable at low temperatures, where the kinetic energy of surface atoms is insufficient to kick out the projectile that stuck to the surface.

2. The next criterion is the one based on an energy calculation and is an application of the criterion Equation 2.1 for the surface tasks: the total translational energy of the projectile atoms should be negative, in order to comply with the sticking criterion. Again, this criterion should be modified if sizeable numbers of atoms at the surface acquire energy close to the threshold energy for sticking.

The sticking coefficient was calculated as a part α of the total number of atoms N that stuck to the surface, N_{st}:

$$\alpha = \frac{N_{st}}{N}. \tag{2.33}$$

Simultaneously with Equation 2.33, a square of the sticking coefficient was calculated and the result was obtained as follows:

$$\alpha = \langle \alpha \rangle \pm \left(\frac{\sqrt{\left(\langle \alpha^2 \rangle - \langle \alpha \rangle^2 \right)}}{\sqrt{M-1}} \right), \tag{2.34}$$

where M is the total number of calculations.

2.9.1 Criterion and the coefficient of surface erosion by energetic cluster impact

At an impact of energetic cluster ions with solid surfaces of targets, the latter experiences heating, melting, and ejection of the matter into gas. The number of ejected atoms is much larger than those that are actually eroded and detached from the surface. The following simple criterion was chosen in [10]: an atom was considered as stuck if its $z_i > 0$ and $v_{zi} > 0$, where z_i is the position of an arbitrary atom on the surface.

If the ejected atom was moving further to a certain position z_0, and if it had velocity component >0, it was accounted as reflected. The sputtering yield Y was calculated as the absolute number of sputtered atoms per cluster atom.

2.9.2   Interaction potentials

Interactions between the particles in rarefied and dense gases, as well as in liquid argon, was modeled by the Lennard-Jones potential (12-6) [40]:

$$U_{LJ}(r) = 4\varepsilon \left[\left(\frac{\sigma}{r} \right)^{12} - \left(\frac{\sigma}{r} \right)^6 \right],$$

(2.35)

where:

ε is the depth of the potential well

σ is the effective atomic (molecular) diameter

In a case where the system contains two kinds of atoms and each kind has a similar type of interaction potential between the species Equation 2.35, with different parameters ε_1, ε_2, then the interaction between the two different atoms can be approximated by a type of potential function similar to Equation 2.35, with the following parameters:

$$\varepsilon_{12} = \sqrt{\varepsilon_1 \varepsilon_2},$$

$$\sigma_{12} = \frac{\sigma_1 + \sigma_2}{2}.$$

(2.36)

Interactions between the particles of most simple metals can be approximated by a Morse potential [31]:

$$U_M(r) = D \left[e^{2\alpha(r_0 - r)} - e^{\alpha(r_0 - r)} \right],$$

(2.37)

where:

D is the depth of the potential

r_0 is the equilibrium position between two interacting particles, the ionic radius

α is the stiffness of the potential function

The Morse potential is convenient for analytical force calculations due to the exponential functions.

In [25], the interaction between an Al_n cluster ion, where n is the number of atoms in cluster, and an Al surface and of a Mo_n ion and a Mo surface were modeled via a potential function that was calculated by an embedded atom method (EAM). It is well known that the cohesion properties of metals are defined by a high density of d-electron states [41]. However, it was soon realized that the details of the density of states (DOS) are not affecting the cohesion energy of an atom in the lattice, which can be determined as a function of a few adjustable parameters:

$$E_b^i = -\left\{ \sum_i \xi^2 e^{\left[-2q(r_{ij}/r_0 - 1)\right]} \right\}^{1/2}, \tag{2.38}$$

where:

ξ is the effective overlap integral

r_0 is the distance between the nearest neighbors

r_{ij} is the interatomic distance

q is a parameter that defines the spatial dependence of the overlap integral

These two parameters are used as adjustable parameters to fit the potential function Equation 2.38 to a large set of experimental data on different properties. The potential Equation 2.38 describes an attractive part of the total potential.

The reflective part of the total atomic energy can be approximated by a Born-Mayer type potential function [42]:

$$E_c = -\sum_i \sum_j A e^{-p(r_{ij}/r_0 - 1)} - \left\{ \sum_j \xi^2 e^{-2q(r_{ij}/r_0 - 1)} \right\}^{1/2}, \tag{2.39}$$

where:

A, ξ are the energy variables

p, q are the adjustable potential parameters, which are obtained by fitting the potential to experimental data including the cohesion energy, the lattice parameter, elastic and shear constants, vacancy, defect energies, and expansion coefficient [43]

If the summation in Equation 2.39 is limited to the nearest neighbors only, then one gets the following expressions for the potential parameters [42]:

$$\xi = \frac{p}{p-q} \frac{E_c}{Z}, \quad A = \frac{q}{p-q} \frac{E_c}{\sqrt{Z}}. \tag{2.40}$$

The analytical expressions for the forces are necessary for MD simulations and can be obtained from Equation 2.39 [42]:

$$F_{ij} = -\frac{2Ap}{r_0} e^{-p(r_{ij}/r_0-1)} + \frac{q\xi}{r_0} \left[\phi_i^{-1/2}(2q) + \phi_j^{-1/2}(2q) \right] e^{-2q(r_{ij}/r_0-1)},$$

$$\phi_k(\lambda) = \sum_l e^{-\lambda(r_{kl}/r_0-1)},$$

$$\vec{F}_i = \sum_{j \neq i} F_{ij} \, \vec{r}_{ij}/r_{ij}, \tag{2.41}$$

where:
   F_{ij} is the absolute values of the interaction forces between atoms i and j
   $\vec{F}_i$ is the total force acting on an atom i

For surface sputtering simulations, the Cu target has been represented by a cylindrical fragment of an FCC structure of 10^5 Cu atoms embedded into the rest of the target and treated by continuum mechanics and linear thermodynamics. The thermal boundary between the two volume parts was organized in such a way that it permitted energy to flow from the impact zone to the bulk and enabled the study of the system's long-term response. The interactions between the copper target atoms were modeled by the EAM and the many-body potential derived from a second-momentum approximation of the tight-binding scheme. The following set of parameters has been used for the potential: $E_c = 3.50$ eV, $Z_{nn} = 12$, $p = 10.08$, $q = 2.56$, where E_c is the lattice cohesion energy, Z_{nn} is the number of nearest neighbors, and p and q are the potential function exponents (Table 2.1).

Table 2.1 Values of the Parameters for Gold and Copper

Material	a, Å	Ec (eV)	A (eV)	ξ (eV)	p	q
Gold	4.079	3.78	0.2061	1.790	10.15	4.13
Copper	3.615	3.50	0.0855	1.224	10.08	2.56

Source: M. Daw and M. Baskes, *Phys Rev Lett*, 50, 1285, 1983.

This choice of parameters has been shown in [42] to describe correctly the room temperature and some high-temperature behavior of bulk Cu in thermal equilibrium. The reflection part of the potential at short distances was not changed because the energies of the impinging clusters that we have studied in this paper were low. Interactions between Ar cluster atoms and Ar and Cu atoms were modeled with a pair-additive Buckingham-type potential. The sputtering yield Y (atoms/ion) is defined as the number of target atoms removed from the surface with one cluster impact. We have obtained this value as a long-time limit of a function $y(t)$ which represents the total number of atoms that crossed a certain control plane at a height z_{cut} above the surface, with z_{cut} taken as a parameter. The value of $z_{cut} = 30$ Å was chosen based on the results presented in Figure 2.5, which indicate that the atoms crossing that plane, well beyond the radius of atomic interaction, will leave the

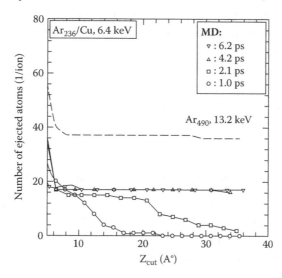

Figure 2.5 Number of ejected Cu target atoms for the Ar_n ($n=236$), $E=6.4$ keV, impact that crossed different z_{cut} planes at different times after a cluster impact. (From Z. Insepov, et al., *Mater Chem Phys* 54, 234–237, 1998.)

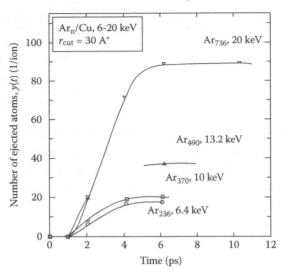

Figure 2.6 Number of ejected Cu target atoms that crossed $z_{cut}=30$ Å plane as a function of time. (From Z. Insepov, et al., *Mater Chem Phys* 54, 234–237, 1998.)

solid [44]. Figure 2.6 shows function $y(t)$ for impacts of Ar clusters of various sizes and energy on a Cu (100) surface for $z_{cut}=30$ Å.

The number of atoms crossing the z_{cut} plane increases with the time elapsed from the beginning of impact and then saturates at a value corresponding to the sputtering yield. These target atoms acquire sufficiently high normal momenta, after the cluster impact, to overcome the surface attraction potential, and are therefore considered as sputtered. The sputtering yield and the angular distribution of the sputtered atoms derived with this method are discussed in Section 9.5.

2.10   Surface Modification Model

To study the surface morphology of a target, we have used a Kuramoto–Sivashinsky (KS) equation [44–46] that was modified by adding a Monte Carlo (MC) term, f_{MC}, to treat the effect of crater formation at the cluster impact:

$$\frac{dh(r,t)}{dt} = \upsilon\nabla^2 h(r,t) - \kappa\nabla^4 h(r,t) + \lambda\nabla^2\left(\nabla h(r,t)\right)^2$$

$$+ f_{MC}\left(\nabla h\right) + \eta(r,t), \tag{2.42}$$

where the first term describes the evaporation and condensation processes of the surface. This term is small for metals and can be neglected in a comparison with the second term. The coefficient υ was set to zero. The second term, reflecting the surface diffusion, can be expressed as $k = D_s \gamma \Omega^2 n_0 / k_B T$, where D_s is the surface diffusion coefficient, γ is the surface tension, Ω is the atomic volume, n_0 is the number of atoms per unit area, k_B is the Boltzmann constant, and T is the surface temperature [45]. The nonlinear term with the coefficient η provides surface current conservation [46]. The Monte Carlo term f_{MC} reflects random events of surface erosion caused by cluster bombardment. The last term in the equation is a white noise that is added to mimic randomly eroded and redeposited atoms.

The modified KS equation has been transformed to a difference equation and computed numerically. The calculations were performed for the cell size $L^2 = 50 \times 50$ nodes, and the periodic boundary conditions were applied along the x- and y-axes. Initially, the rough surface was represented by a single hill placed in the center of the computational cell. After the creation of a crater at a random position, we computed the KS equation typically for 1000 time steps, with different diffusion coefficients.

2.11   Summary of Chapter 2

1. A new criterion of cluster identification in a dense gas was developed that is based on the negativity of the total internal energy of a spatially confined group of atoms.

2. Molecular models of formation, excitation, stabilization, and evaporation of clusters in rarefied gases were developed.

3. A simple model of electronic excitation of a cluster was discussed that can be applied to simulate the energetic cluster ion impacts on a surface.

4. A TST model of transition complex for "cold cluster evaporation" was developed that was supplemented by a compensation potential method.

5. A multizone surface model was developed that was based on the principle incomplete description of "uninteresting" degrees of freedom and that facilitates simulation of strong shock waves generated at energetic cluster ion impacts on a surface.

References

1. D. W. Heermann, *Computer Simulation Methods in Theoretical Physics*, 2nd edition, Springer: Berlin, 1990.

2. H. Muller-Krumbhaar, Simulation of small systems, in K. Binder (ed.) *Monte Carlo Method in Statistical Physics*, vol. 7, Springer: Berlin, 1986.

3. E. E. Polymeropoulos and J. Brinkman, Molecular dynamics study of the formation of argon clusters in the compressed gas, *Chem Phys Lett* 92, 59–63, 1982.

4. E. E. Polymeropoulos, P. Bopp, J. Brinkman, L. Jansen, and R. Block, Molecular-dynamics simulations in systems of rare gases using Axilrod-Teller and exchange three-atom interactions, *Phys Rev A* 31, 3565–3569, 1985.

5. Z. Insepov and S. V. Zheludkov, About clusters near the coexistence line, *High Temp* 25, 607–609, 1987.

6. Z. Insepov, E. Karataev, and G. E. Norman, The kinetics of homogeneous condensation behind the shock front, *Z Phys D* 20, 449–451, 1991.

7. Z. Insepov and E. Karataev, Molecular dynamics calculation of growth and decay rate constants behind the shock front, *Math Modeling* 5, 48–56, 1993.

8. R. D. Etters and J. Kaelberer, Thermodynamics properties of small aggregates of rare-gas atoms, *Phys Rev A* 11, 1068–1079, 1975.

9. M. H. Shapiro and T. A. Tombrello, Simulation of core excitation during cluster impacts, *Phys Rev Lett* 68, 1613–1615, 1992.

10. Z. Insepov, M. Sosnowski, and I. Yamada, Molecular dynamics simulation of metal surface sputtering by energetic rare-gas cluster impact, in I. Yamada et al., (ed.), *Laser and Ion Beam Modification of Materials*, in *Proceedings of the IUMRS-ICAM-93*, pp. 111–118, Tokyo, Japan, 1993.

11. R. Beuhler and L. Friedman, Larger cluster ion impact phenomena, *Chem Rev* 86, 521–537, 1986.

12. J. M. Soler, J. J. Saenz, N. Garcia, and O. Echt, The effect of ionization on magic numbers of rare-gas clusters, *Chem Phys Lett* 109, 71–75, 1984.

13. D. Fenyö, B. U. R. Sundqvist, B. R. Karlsson, and R. E. Johnson, Molecular dynamics study of electronic sputtering large organic molecules, *Phys Rev B* 42, 1895–1902, 1990.

14. A. Polian, P. Loubeyre, and N. Boccara, *Simple Molecular Systems at Very High Density*, NATO ASI Series 4, pp. 690–696, Plenum: New York, 1989.

15. I. M. Sobol, A Primer for the Monte Carlo Method, CRC Press: Boca Raton, FL, 1994.

16. B. F. Gordiets, A. I. Osiov, and L. A. Shelepin, *Kinetic Processes in Gases and Molecular Lasers*, Taylor & Francis: London, 1988.

17. A. V. Krestinin, Simple model of non-isothermal homogeneous nucleation in gases (In Russian), *J Chem Phys* 5, 240–249, 1986.

18. K. A. Holbrook, M. J. Pilling, S. H. Robertson, and P. J. Robinson, *Unimolecular Reactions*, vol. 93, Wiley: New York, 1996.

19. S.-N. Yang and T.-M. Lu, The sticking coefficient of Ar on small Ar clusters, *Solid State Commun* 61, 351–354, 1987.

20. I. V. Kazakova, Numerical modeling of condensation processes from gas phase, Ph.D. thesis, Novosibirsk, 1994.

21. Z. Insepov and E. Karataev, Molecular dynamics simulation of infrequent events: Evaporation from cold metallic clusters, NATO ASI Series, *Phys Chem Finite System* 1, 423–427, 1992.

22. A. F. Voter, A method for accelerating the molecular dynamics simulation of infrequent events, *J Chem Phys* 106, 4665, 1997.

23. E. K. Grimmelman, J. C. Tully, and E. Helfand, Molecular dynamics of infrequent events: Thermal desorption of xenon from a platinum surface, *J Chem Phys* 74, 5300–5310, 1981.

24. Z. Insepov and A. Zhankadamova, Molecular dynamics calculations of the sticking coefficient of gases to surfaces, *Z Phys D* 20, 145–146, 1991.

25. H. Haberland, Z. Insepov, and M. Moseler, This film grow by energetic cluster impact (ECI): Comparison between experiment and molecular dynamics simulation, *Mater Sci Eng B* 19, 31–36, 1993.

26. H. Haberland, Z. Insepov, and M. Moseler, Molecular dynamics simulation of thin film formation by ECI, *Z Phys D* 26, 229–231, 1993.

27. H. Haberland, Z. Insepov, and M. Moseler, Molecular dynamics simulation of cluster deposition, *Nucl Instrum Methods Phys Res B* 89, 419–425, 1993.

28. M. Ishii, G. Sugahara, Z. Insepov, M. Kumagai, G. H. Takaoka, J. Northby, and I. Yamada, Molecular dynamics simulation for cluster bombardment of solid surface, in *Proceedings of the 40th Spring Meet, Japan Soc Appl Phys* 2, 588, IOP, 1993.

29. Z. Insepov, M. Ishii, G. Sugahara, M. Kumagai, G. H. Takaoka, J. Northby, and I. Yamada, Molecular dynamics simulation for sputtering of gold surface by an energetic argon cluster, in *Proceedings of the 40th Spring Meet, Japan Soc Appl Phys* 2, 588, IOP, 1993.

30. F. O. Goodman and H. Y. Wachman, *Dynamics of Gas-Surface Scattering*, Elsevier: Waltham, MA, 1976.

31. I. Yamada, J. Matsuo, Z. Insepov, D. Takeuchi, M. Akizuki, and N. Toyoda, *J Vac Sci Technol A* 14, 781, 1996.

32. Z. Insepov and I. Yamada, Molecular Dynamics simulation of cluster ion bombardment of solid surfaces, *Nucl Instrum Methods Phys Res B* 99, 248, 1995.

33. M. Moseler, J. Nordiek, and H. Haberland, Reduction of the reflected pressure wave in molecular dynamics simulation of energetic particle-solid collisions, *Phys Rev B* 56, 15439, 1997.

34. Z. Insepov, M. Sosnowski, and I. Yamada, Simulation of cluster impacts on silicon surface, *Nucl Instrum Methods Phys Res B* 127, 269, 1997.

35. Z. Insepov, R. Manory, J. Matsuo, and I. Yamada, Proposal for a hardness measurement technique without indentor by gas-cluster-beam bombardment, *Phys Rev B* 61, 11605, 2000.

36. Y. B. Zel'dovich and Y. P. Raiser, *Physics of Shock Waves and High-Temperature Hydrodynamic Phenomena*, p. 653, Academic Press: New York, 1967.

37. L. D. Landau and E. M. Lifshitz, *Theory of Elasticity*, vol. 7, Pergamon: Oxford, 1986.

38. D. R. Lide (ed.), *Handbook of Chemistry and Physics*, p. 4–100, CRC Press: London, 1993.

39. Z. Insepov and E. Karataev, Molecular dynamics model of rate constant calculation of evaporation from cold metallic clusters, NATO ASI Series, *Phys Chem Finite System* 1, 423–427, 1992.

40. F. F. Abraham, Computational statistical mechanics: Methodology, applications and supercomputing, *Adv Phys* 35, 1–111, 1986.

41. C. Kittel, *Introduction to Solid State Physics*, 6th edition, pp. 143–146, Wiley: New York, 1986.

42. V. Rosato, M. Guillope, and B. Legrand, Thermodynamical and structural properties of fcc transition metals using a simple tight-binding model, *Phil Mag A* 59, 321–336, 1989.

43. M. Daw and M. Baskes, Semiempirical, quantum mechanical calculation of hydrogen embrittlement in metals, *Phys Rev Lett* 50, 1285, 1983.

44. Z. Insepov, I. Yamada, and M. Sosnowski, Sputtering and smoothing of metal surfaces with energetic gas cluster beams, *Mater Chem Phys* 54, 234–237, 1998.

45. W. W. Mullins, Theory of thermal grooving, *J Appl Phys* 28, 333, 1957.

46. Z.-W. Lai and S. Das Sharma, Kinetic growth with surface relaxation: Continuum versus atomistic models, *Phys Rev Lett* 66, 2348, 1991.

CHAPTER 3

Molecular Dynamics Method

3.1 Introduction

Molecular dynamics (MD) has been widely used for calculating the structure and equilibrium properties of various systems [1–5]. Only recently, however, has MD been used to study the kinetics of phase transitions. In this chapter we discuss the simulation of the kinetics of first-order phase transitions. Initial and boundary conditions are analyzed, and various simulation ensembles and algorithms are described.

The kinetics of phase transitions is important for understanding what happens during realistic physical processes. According to our calculations, it was determined that employing a canonical ensemble (*NVT*) gives more accurate characteristics of these processes compared to a microcanonical ensemble (*NVE*).* In a microcanonical ensemble, oscillations of the condensation–evaporation sequences occur due to the heating of the system. For modeling swift collisions, the most computationally inexpensive method is to use an isobaric ensemble.

In the study of the nonequilibrium properties of the system, averaging is not so trivial. Therefore, a method for averaging is also addressed in this book. We propose averaging over an ensemble of statistically independent initial values of the coordinates and velocities of the system. Choice of the system size for simulations of nonequilibrium properties is also important. According to the analysis in this book, the necessary system size may not be larger than several tens or hundreds of particles in the basic cell.

3.2 Isothermal Ensemble of Molecular Dynamics

The MD method initially was formulated as a solution of the equations of motion of a system of N interacting particles in a volume V and with total energy E. This is a *microcanonical*, or *NVE*, *ensemble* of statistical mechanics [1–3].

---

* In NVT and NVE, N, V, T, and E are constants, where N is the number of atoms, V is the system volume, T is temperature, and E is the energy.

By carefully choosing all coordinates and moments of the atoms in the system, researchers assumed they could set up a given total energy. In such a setup, after the evolution of the system started, it would move along a trajectory in phase space with a constant total energy.

As our calculations showed [4,5], however, using a microcanonical ensemble led to oscillation of the condensation–evaporation sequences, caused by the release of the latent heat of condensation and heating up the system. Such behavior is not typical for most simple systems and is a consequence of the isolation of the system in the microcanonical NVE ensemble.

A typical characteristic of any real physical process is the thermal and mass exchange of the molecular system with the rest of the environment, which stabilizes the system temperature, pressure, and the density. The energy of the endothermic and exothermic processes is exchanged with a thermostat. The total energy of the subsystem is no longer an integral of motion. The temperature T in such a subsystem is a variable fluctuating about an average value. By using special algorithms, one can define the level of fluctuations by the correct value of the thermal specific heat. Such an ensemble is called a *canonical*, or *isothermal, NVT ensemble* [1,3,6,7].

The simplest way to set up an isothermal system is to use a scaling factor by which all velocities in the subsystem are multiplied at each time step of integration of the equations of motion [3]:

$$v_i' = k \cdot v_i,$$

$$k = \left[\frac{(3N-4)k_B T_{ref}}{\sum_i m v_i^2} \right], \tag{3.1}$$

where T_{ref} is the temperature that is set up before the calculation and that the system's average kinetic energy per atom/molecule (a thermodynamic definition of temperature) approaches, after the system passes the initial nonequilibrium stage. Usually, instead of $3N-4$, Equation 3.1 uses the factor $3N$. If the average temperature deviates from T_{ref}, it can be adjusted by an increase (or decrease) of the value of T_{ref}.

The relevance of the choice of a thermostat method can be exemplified by the adsorption of atoms on a solid surface [8]. In this case, velocity rescaling is reasonable because all the adsorbed atoms on the surface experience similar fluctuations caused by interactions with substrate atoms. If, however, the subsystem has no direct contact with another system (such as a surface or a buffer gas), a different choice of thermostat method is needed.

3.2.1 Stochastic thermostats

In [4,5], we introduced an isothermal ensemble that used a stochastic process at the thermal walls. In this ensemble, the system energy is assumed to fluctuate as a result of an energy exchange with the substrate. Periodic boundary conditions are modified as follows: if an atom leaves the system, the periodic image of the atom is entered in the system with a Maxwellian velocity distribution corresponding to a reference temperature. For a short run, this method can efficiently keep the desired reference temperature even if the system is small; for example, only 109 atoms were used in the work reported in [9,10]. For long runs, however, the method is not efficient because the atoms leaving the system are the fastest, and hence the system cools down to a greater extent than it is heated by the introduction of new atoms at an average temperature.

3.2.2 Deterministic thermostats

Nose proposed a new method of canonical ensemble in which an artificial mass was added to the Hamiltonian [7]. A limitation of this deterministic approach, however, is that the fictitious particle has to be assigned a position, a velocity that has no clear physical meaning.

3.3 Brownian Dynamics

Allen proposed a numerical approach using Brownian dynamics [11]. In this approach a small volume (three-dimensional) of the subsystem interacts with the thermostat. Specifically, the approach mimics the interaction of an atomic system, for example, a vapor, with a noncondensing buffer gas. The equations of motion contain additional friction and random forces:

$$\vec{F}_i = \vec{F}_{\text{pot},i} - \gamma\vec{v}_i + \vec{R}_i \quad (i = 1, 2, ..., N), \tag{3.2}$$

where:
  γ is the friction coefficient
  v_i is the atomic velocity
  R is the random force

The module of the random force is assumed to be distributed over a Gaussian law and is directed uniformly in the space. For simplification, the random force is assumed to be δ-correlated and can be obtained from the following equations [1,8]:

$$\langle R_{\alpha i} \rangle = 0, \quad \alpha,\beta = x,y,z; \quad i,j = 1,2,...,N.$$

$$\langle R_{\alpha i}(t) R_{\beta j}(t') \rangle = \frac{2k_B T \gamma \delta(t-t') \delta_{\alpha\beta} \delta_{ij}}{m}, \tag{3.3}$$

where:
  $\delta(t)$ is a δ-function of Dirac
  m is an atomic mass

Since in MD the equations are integrated with a time increment Δt and a random force according to Equations 3.2 and 3.3 will be transformed into the following:

$$\langle R_{\alpha i}^2 \rangle = \frac{2k_B T \gamma}{(\Delta t \cdot m)}, \tag{3.4}$$

where $\langle ... \rangle$ means averaging over time.

A significant deviation exists over the value of the friction constant γ. Tully et al. [12,13] studied the motion of an adsorbed particle on a solid surface by solving a Langevin equation and approximating the friction constant as follows:

$$g = \frac{pw_D^{\ s}}{6},$$

where:
  $\omega_D^{\ s}$ $= 1/2 \, (k_B\theta)/\hbar$ is the Debye frequency of the surface atoms
  θ is the Debye temperature for volume atoms
  $\hbar$ is Plank's constant

Applicability of Brownian dynamics is related to the mass ratio of the condensing gas to the mass of the buffer gas. However, a recent study [14] showed that this classical approach is not mandatory. It was realized that a particle placed in a medium of similar particles behaves as does a Brownian particle at times much larger than the collision times, and therefore one can introduce a friction force.

Another necessary condition of Brownian dynamics is the high frequency of collision of the condensing atoms with the buffer gas. If no buffer gas exists, the collision frequency of the particles between each other should be much larger than the characteristic times of growth and evaporation of the cluster τ_{cl}:

$$v = n\sigma v \gg \frac{1}{\tau_{cl}}, \tag{3.5}$$

where:

    n is the density of particles

    σ is the cross section of interaction between the particles of the condensing gas

    v is the relative velocity of the colliding particles (atoms or molecules)

In other words, the dynamics at the smaller time scales $\sim v^{-1}$ is insufficient to be described by Brownian motion. The absolute value (modulus) is distributed uniformly within the interval $\left[-\sqrt{6\gamma k_B T/\Delta t m}, +\sqrt{6\gamma k_B T/\Delta t m}\right]$, and the direction is isotropic. The friction coefficient γ is the only variable that relates this method with the parameters of the modeling system such as mass, density, and temperature. For an estimate of this parameter, one can use a gas dynamic expression for the cross section in a rarefied gas:

$$\gamma = \rho_g <\sigma v>. \tag{3.6}$$

For the friction coefficient in liquid, a more appropriate formula is the Stokes formula:

$$\gamma = 6\pi R \cdot \eta, \tag{3.7}$$

where:

    η is the dynamic viscosity

    R is the cluster radius

In simulating the cluster formation process, care should be taken that the random forces imitating the collisions with other particles are not acting on the atoms inside the cluster. To this end, one can rescale the friction coefficient, making it smaller at larger cluster sizes, for example, by dividing by an average cluster size:

$$\gamma' = \frac{\gamma}{\langle k(t) \rangle^{1/3}}, \tag{3.8}$$

where $\langle k(t)... \rangle$ is the average cluster size at the time instant t.

The friction coefficient can be obtained from the Stokes formula given in Equation 3.7. One can then assume that the total force acts on all particles, with the value rescaled according to the cluster size, as in Equation 3.8. The physical meaning of the friction and random forces is that they control the contact of the system with the thermostat and, by doing so, generate a correct level of thermodynamic fluctuations of the temperature and other variables of the system. The main variable of the canonical ensemble is temperature; for example, its fluctuations determine the specific heat capacity [15].

In [16], we selected the friction constant using an experimental value for the specific heat C_V. The specific heat capacity for liquid argon was calculated by using MD, and the atomic interactions were approximated by using a Lennard-Jones potential. The density and temperature were $rs3 = 0.75$ and $k_B T/e = 0.91$ (e and s are the parameters of the potential). The specific heat was calculated by using the expression

$$\frac{C_V}{N k_B} = 3/2 + N^{-1}(k_B T)^{-2}\left(\langle U_N^2 \rangle - \langle U_N \rangle^2 \right), \tag{3.9}$$

where:

N is the total number of the particles in the basic cell,
$\langle U_N^2 \rangle$ is the average of the square of the total potential energy
$\langle U_N \rangle^2$ is the square of the average energy

The calculations were carried out for three values of γ, namely, 7.5, 15, and 22.5, in units of σ^{-1} $(\varepsilon/m)^{1/2}$. The values of the specific heat capacities were consequently 2.29, 2.31, and 2.53, in units of k_B. An experimental value of the specific heat capacity is 2.29 [17]. Therefore, the best value of γ obtained from the experimental specific heat is 7.5 (Figure 3.1).

We note that the Stokes formula Equation 3.7 in this case gives the value $\gamma = 15.0$. Calculation of the specific heat using velocity

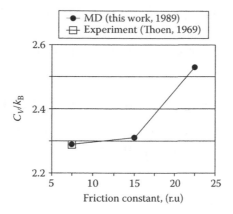

Figure 3.1 Calculation of the friction coefficient by comparison of MD with experiment. Specific heat obtained by MD in comparison with the experimental value for liquid argon. (Data from Z. Insepov, et al., Canonical ensemble by molecular dynamics, in *MECO-18, 18th Seminar of the Middle-European Cooperation in Statistical Physics*, Duisburg, Germany, p. 71, 1991; J. Thoen, et al., *Physica* 45, 339–345, 1969.)

scaling gives a value $C_V/k_B = 2.96$, which is much larger than the experimental value [17].

A similar model was developed for studying the kinetics of recrystallization in liquid argon, as well as for studying the Soret effect [18]. For modeling adsorption layers on solid surfaces by Brownian dynamics, the friction coefficient was selected based on the work in [19], where it was assumed that relaxation of the atoms was achieved as a result of an energy exchange between the adsorbed atoms and acoustic phonons in the surface. The friction coefficient was calculated to be

$$\gamma = 9\pi m_a \frac{\omega^4(0)}{\omega_s^3}, \tag{3.10}$$

where m_a and m_s are the atomic masses of adsorbed and surfaces atoms, respectively

$$w(0) = \alpha \left(\frac{2E_a}{m_a} \right)^{1/2}, \quad a = \frac{3.6}{a},$$

where:

      a is the lattice parameter

E_a is the energy of activation of the diffusion previously calculated by MD

$\omega_s \approx 2\,\omega(0)$ is the frequency of the surface phonons.

3.4   Isobaric Ensemble for Modeling Swift Collisions

MD simulations involve a large number of particles and therefore are computationally expensive. Researchers should choose between the size of the simulation system and the length of simulations in real time. It is not reasonable, for instance, to calculate a system with a billion atoms for several hundred nanoseconds. In some cases, however, the simulation should be long or the system sufficiently large to correspond to a realistic physical phenomenon. Here we describe one such case related to ion surface interaction that is important for the semiconductor industry and nuclear materials science and engineering.

The collision of fast moving gas particles (atoms or molecules) bombarding a target's surface with energies of 1–100 eV/atom creates a strong spherical elastic and/or inelastic pulse wave that propagates into the bulk with speed u. During the interaction time interval τ, the wave encompasses the region of the surface with a radius of $R \sim u\tau$ and the number of atoms inside the region $N \sim \rho_0 V = \rho_0\, u^3\tau^3$, where ρ_0 is the density of the solid. Using the parameters $u \sim 1\,\text{km/s}$, $\tau \sim 1\text{–}10\,\text{ps}$, and $\rho_0 \sim 10^{22}\,1/\text{cm}^2$, we obtain $N \sim 10^3\text{–}10^4$. Therefore, simulating the interaction of a fast ion with a surface needs both a large number of atoms in the sample and a large total number of samples, in order to make a proper averaging over the samples.

With a small number of atoms, less than 104, the elastic waves were able to be reflected back to the collision zone in the center of the spot and destroy the dynamics of the atomic motion. As a result, the simulation was not adequate. In this case, the results also depend on the shape of the sample, because of the cylindrical or spherical geometry of the elastic waves.

We therefore devised a simulation method that enables a variable MD basic cell, in principle, to completely remove reflected elastic and thermal waves, thus enabling more economic calculations of the collision [20–22]. Pressure inside the system was calculated by using a virial formula [1–3]:

$$\frac{P}{nk_{B}T}=1-\frac{\left\langle \sum_{i<j}r_{ij}\,\partial U_{ij}/\partial r_{ij}\right\rangle}{6Nk_{B}T}-\frac{n\int_{r_{0}}^{\infty}r\,dU/dr\,g(r)dr}{6k_{B}T}, \tag{3.11}$$

where r_0 is the cut-off radius of the interaction potential for the surface particles. Averaging was carried out over all particles in the basic cell. According to the isothermal compressibility β, the pressure variation dP in the basic cell leads to a volume variation

$$\frac{dV}{V}=-\beta dP, \tag{3.12}$$

where the compressibility values can be obtained from the cold compression curves, since the system cannot noticeably heat up within a very short time of collision $\leq\tau$. The relative volume change Equation 3.12 can then be used to obtain the elongation of the linear system size. This method was used to obtain the volume change for a hemispherical system [23,24]; the results showed that the elastic waves can be significantly suppressed at a system size of $N\approx6000$.

3.5 Averaging over the Initial States

The MD method mainly has been used to calculate equilibrium properties of matter. The nonequilibrium part of the system evolution was usually discarded as unphysical [14,25–27]. In [3], for example, averaging over an ensemble of various system properties that were dependent on the dynamic variables was replaced by averaging over time on the equilibrium part of the system evolutions:

$$\langle A\rangle_{NVT}=\lim_{T\to\infty}\frac{1}{T}\int_{0}^{T}A\left(\vec{r}^{N}(t),\vec{v}^{N}(t)\right)dt, \tag{3.13}$$

where:

   A is an arbitrary function of coordinates $\vec{r}^{N}$ and velocities $\vec{v}^{N}$ of a system of N particles

   T is the period of averaging

The challenge with MD, then, is to determine how to use this averaging Equation 3.13 to study the nonequilibrium stage of the system evolution. Obviously, the ergodic theorem Equation 3.13 will be breached, and the MD method should be supplemented with a new rule of averaging of the functions of dynamic variables.

To this end, we present a method for averaging over an ensemble of statistically independent initial values of coordinates and velocities of the system. This method was first applied for solving problems of weakly nonequilibrium plasma [14]. For the present study, the ensemble was fabricated by sampling all the coordinates and velocities of the system according to a uniform spatial probability density and Maxwellian probability of the velocities at the same temperature. We call each start of the simulation from an individual initial state (including all coordinates and velocities) a "realization." If the number of such realizations is M, then the averaging over the ensemble will be determined as follows:

$$\langle A(t) \rangle_{NVT} = \lim_{M \to \infty} \frac{1}{M} \sum_{i=1}^{M} A_i\left(\vec{r}^N(t), \vec{v}^N(t)\right), \tag{3.14}$$

although, in reality, the number of independent initial states M typically is selected between 10 and several hundreds.

3.6   Selection of the Number of Particles in the Basic Cell

Generally, the thermodynamics and statistical mechanics laws are preconditioned because of the large number of particles (on the order of Avogadro's number) and the incompleteness of the system description. As shown by Valuev et al. [14] in their discussion of the theory of MD, however, the statistical physics laws emerge in MD as a result of a phenomenon called *intermixing*. Therefore, the choice of the system size is determined by other reasons.

As an example, in using MD to study equilibrium systems, the system size should be larger than the correlation length. The specific system size can be determined by the required accuracy for the problem and can be obtained empirically by calculation. On the other hand, calculations of temporal—that is, nonequilibrium—characteristics should be restricted to those trajectories that are not leaving the system during the calculation interval [14]:

$$L^2 \gg v_T^2 \tau_v t_p, \tag{3.15}$$

where:

    L is the system size

    t_v is the correlation time of velocity vectors

    v_T is the thermal speed

As pointed out in [14], for calculating nonequilibrium properties in dense systems, the necessary system size may not be larger than several tens or hundreds of particles.

3.7   Calculation Technique

The MD method involves the numerical solution of Newton's or Langevin's equations of motion for all N particles included in the basic simulation cell. The equations of motion were integrated by using Verlet's (see Section 1.7) technique, with a time increment that is dependent on temperature and on the relative velocity of approaching particles (for the tasks where particles interact with surfaces).

If the simulated system is small ($N \leq 100$), usually no problems occur with acceleration of the calculations, although in traditional MD the calculation time depends as $\sim N^2$ on the number of atoms [3].

If the system size increases, however, one needs special algorithms to speed the force calculation, which takes about 90% of the calculation time, and to replace the square dependence (N^2) with a linear $\sim N$ or a logarithmic one ($\sim N \ln N$) [28]. For the acceleration, a Verlet list procedure [1] is used in this chapter because of its simplicity; it has a speedup comparable to that of the Hockney–Eastwood method of linked lists for small systems where $N \leq 1000$. The cut-off radius of Verlet's sphere was selected as $r_0 = 2.6\sigma$, where σ is the atomic diameter. In this method, the numbers of all neighbors inside the external Verlet sphere are stored in the Verlet list. These numbers are then used at each time step to calculate the neighbor's forces. The atoms inside the internal Verlet sphere with radius $r_0 - dr$, where $dr \approx 0.5\sigma$, are taken into account at each time step. The Verlet list is renewed once in 10–20 time steps, depending on the average atomic velocity.

For simulating bulk systems, a nearest-image technique is used [1,2], where the atoms near the border interact with the copies of the atoms placed in neighboring cells. For simulating the kinetic processes of cluster formation, the average density of clusters of all sizes depending on time is as follows:

$$\overline{n_k} = \left\langle \frac{N_k}{\sum\limits_{k \geq 2} N_k} \right\rangle, \tag{3.16}$$

where N_k is the number of clusters of k-atoms at a time instant t and the radial pair-distribution function of atoms in 3D and 2D geometries is

$$g_{3D}(r) = \frac{\Delta N(r)}{4\pi r^2} dr, \tag{3.17}$$

$$g_{2D}(r) = \frac{\Delta N(r)}{2\pi r} dr' \tag{3.18}$$

where $\Delta N(r)$ is the number of atoms inside a spherical volume in 3D or inside a 2D circular area.

A Fourier transform of the pair-distribution function gives the structural factor, which can be calculated directly from the distances between the atoms r_{ij} in the system [26,27]:

$$S(k_\lambda) = 1 + \left(\frac{4}{\pi}\right) N \sum_{i<j} \frac{\sin(k_\lambda r_{ij})}{k_\lambda r_{ij}}, \tag{3.19}$$

where $k\lambda$ is the wave number.

For collisions of clusters with surfaces, the traditional definition of temperature as an average kinetic energy per atoms becomes inadequate because the average kinetic energy will be defined mostly by the acceleration voltage, not by the thermal energy. At an impact of the cluster with the surface, the acceleration energy will be converted into a radial (hydrodynamic) collective motion of the atoms upon collision. Therefore, for such impact processes, we introduce temperature as an average energy per tangential degree of freedom [23]:

$$T_{cl} = \sum_{i=1}^{N_{cl}} \frac{m_i v_{it}^2}{k_B (3N-6)}, \tag{3.20}$$

where m_i and N_{cl} are the mass and the number of atoms in the clusters, respectively. If the new definition of temperature is used, the temperature map shows a spherical symmetry at an impact that correctly reflects the spherical geometry of the collision. Therefore, the temperature is calculated separately for each spherical layer with the central point located in the collision spot at the surface of the target. Apart from that, within the collision process one can calculate the average pressure in the central collision zone for which the virial formula Equation 3.11 was used.

3.8   Summary of Chapter 3

1. The MD method was modified for simulating an NVT ensemble. For this purpose, the boundaries of the basic cell were replaced with "stochastic" boundaries or "thermal" walls. The atoms reflected from such walls were assigned a velocity modulus obtained from a Maxwell distribution at a given temperature.

2. A Brownian dynamics method was used to calculate the cluster kinetic formation in dense gases, in liquids, and inside the adsorption layers on the solid surface. In this approach, the friction constant in the Langevin equation was determined from physical principles.

3. The MD method was supplemented with an algorithm that enables modeling the impact of highly accelerated clusters (with energy up to 100 eV/atom) on a surface of the target.

4. A new averaging procedure over the statistical independent initial state was presented that can be applied to study the kinetic characteristics of cluster formation.

References

1. F. F. Abraham, Computational statistical mechanics: Methodology, applications and supercomputing, *Adv Phys* 35, 1–111, 1986.

2. M. P. Allen and D. J. Tildesley, *Computer Simulation of Liquids*, p. 385, Oxford University Press: Oxford, 1987.

3. D. W. Heermann, *Computer Simulation Methods in Theoretical Physics*, 2nd edn, Springer: New York, 1990.

4. Z. Insepov and S. Zheludkov, On the lack of clusters near the gas-liquid saturation line (in Russian), *High Temperature* 25, 605–609, 1987.

5. Z. Insepov and S. Zheludkov, Molecular dynamics modeling of the kinetics of cluster formation in supersaturated vapor (in Russian), *J Phys Chem* 61, 1109–1111, 1987.

6. V. Valuev, G. Norman, and V. Yu. Podlipchuk, Method of molecular dynamics, in *Mathematical Modeling, Physics-Chemical Properties of Matter* (in Russian), Nauka: Moscow, 1989.

7. S. Nose, A molecular dynamics method for simulation in the canonical ensemble, *Mol Phys* 52, 255–268, 1984.

8. F. F. Abraham, Phases of two-dimensional matter, *Phys Rep* 80, 339–374, 1981.

9. Z. Insepov and E. Karataev, Molecular dynamics simulation of infrequent events: Evaporation from cold metallic clusters, *Phys Chem Finite System*, NATO ASI Series, 374, 423–427, 1992.

10. Z. Insepov and E. Karataev, Molecular dynamics model of rate constant calculation of evaporation from cold metallic clusters, *Tech Phys Lett* 17, 36–40, 1991.

11. M. P. Allen, Algorithms for Brownian dynamics, *Mol Phys* 47(3), 599–601, 1982.

12. J. C. Tully, Dynamics of gas-solid interaction: 3D general Langevin model used for fcc and bcc surfaces, *J Chem Phys* 73, 1975–1986, 1980.

13. M. Shugard, J. C. Tully, and A. Nitzan, Dynamics of gas-surface interaction: Calculation of energy transfer and sticking, *J Chem Phys* 66(6), 2534–2544, 1980.

14. A. A. Valuev, G. E. Norman, and V. Yu. Podlipchuk, Method of molecular dynamics: Theory and applications (in Russian), in A. A. Samarskii and N. N. Kalitkin (eds), *Mathematical Modeling*, pp. 5–40, Nauka: Moscow, 1989.

15. Z. Insepov and V. Zheludkov, Kinetics of clustering on the surface, in *Proceedings of Workshop Nucleation, Clusters, Fractals*, Serrahn, Germany, pp. 102–110, 1991.

16. Z. Insepov, B. Z. Kabdiev, and A. A. Valuev, Canonical ensemble by molecular dynamics, in *MECO-18, 18th Seminar of the Middle-European Cooperation in Statistical Physics*, Duisburg, Germany, p. 71, 1991.

17. J. Thoen, E. Vangeel, and W. van Dael, Heat capacity of liquid argon, *Physica* 45, 339–345, 1969.

18. Z. Insepov and B. Z. Kabdiev, Molecular dynamics simulation of pulsed recrystallization of crystal surface, in *Rostocker Physics Manuscripts, Proceedings of the Workshop Nucleation-Clusters-Fractals*, Serrahn, Germany, pp. 111–112, 1991.

19. V. P. Zhdanov, Some aspects of the dynamics of the motion of adsorbed particles, *Kinetic Catal* 28(1), 247–250, 1987.

20. H. Haberland, Z. Insepov, M. Karrais, M. Mall, M. Moseler, and Y. Thurner, Thin film growth by energetic cluster impact (ECI): Comparison between experiment and molecular dynamics simulations, *Mater Sci Eng B* 19(2), 31–36, 1993.

21. H. Haberland, Z. Insepov, and M. Moseler, Molecular dynamics simulations of thin film formation by ECI, *Z Phys D* 26, 229–231, 1993.

22. H. Haberland, Z. Insepov, and M. Moseler, Molecular dynamics simulations of cluster deposition, *Nucl Instrum Methods* 89, 419–425, 1993.

23. Z. Insepov, M. Sosnowski, and I. Yamada, Molecular-dynamics simulation of metal surface sputtering by energetic rare-gas cluster impact, in *Proceedings of the IU MRS-ICAM-93*, pp. 1–8, Tokyo 1993.

24. H. J. C. Berendsen, J. P. M. Postma, W. F. van Gunsteren, A. DiNola, and J. R. Haak, Molecular-dynamics with coupling to an external bath, *J Chem Phys* 81, 3684–3690, 1984.

25. A. Rahman, Correlation in the motion of atoms in liquid argon, *Phys Rev A* 136, A406–A411, 1964.

26. V. A. Polukhin, V. F. Ukhov, and M. M. Dzugutov, *Computer Modeling of the Dynamics and Structure of Liquid Metals* (in Russian), Nauka: Moscow, 1981.

27. A. N. Lagar'kov and V. M. Sergeev, Molecular dynamics method in statistical physics, *Sov Phys Uspekhi* 21, 566–588, 1964.

28. R. W. Hockney and J. W. Eastwood, *Computer Simulation Using Particles*, Taylor & Francis: New York, 1988.

CHAPTER 4

Kinetics of Cluster Formation in Dense Gases

4.1 Introduction

All existing theories of kinetic processes were created for rarefied gases where the kinetics is controlled by binary collisions. The present chapter is devoted to calculating the kinetics of cluster formations in both nonsaturated and supersaturated dense gases. Specific attention was given to developing criteria of cluster recognition in dense gases and for selecting a proper ensemble of a system experiencing a phase transformation.

To achieve cluster formation in a nonsaturated dense gas, N particles with Lennard-Jones interaction were placed in a cubic cell, with edges L and periodic boundary conditions. The equations-of-motion were solved by a Verlet algorithm in a microcanonical ensemble. The total relative energy of the chosen atom interacting within the Verlet's sphere about the atom was tested to verify that a gas atom belonged to a cluster. It was determined that the pressure in a nonsaturated dense gas decreases with the formation of clusters.

The kinetics of cluster formation in a dense supersaturated vapor was studied in microcanonical (NVE) and canonical (NVT) ensembles.* Oscillations of kinetic and potential energies were observed in NVE ensemble calculations. A qualitative agreement with spinodal decay theory was obtained.

4.2 Cluster Formation in Nonsaturated Dense Gas

In this chapter the criterion of cluster identification developed in Chapter 2 will be used for studying cluster formation in a dense nonsaturated gas.

The system under consideration is represented by a cube basic cell, with edges L, in which N particles were placed that interact

---

* In NVT and NVE, N, V, T, and E are constants, where N is the number of atoms, V is the system volume, T is temperature, and E is the energy.

via a Lennard-Jones model interatomic potential [1,2]. The initial positions and velocities of the system were selected randomly in accordance with the uniform probability density function and the Maxwell distribution function [3].

Periodic boundary conditions were introduced, to remove system boundary effects. The interaction potential was cut off at the cut-off radius of 2.5 σ. The equations of motion were solved by a Verlet algorithm (see Equation 1.18) in a microcanonical ensemble, with N, V, E constant. The desired temperature was provided by a scaling procedure within several tens of time steps. The atomic velocities were multiplied by a factor

$$k = \sqrt{\frac{3/2\,Nk_BT}{\sum_{i=1}^{N} mv_i^2/2}}, \tag{4.1}$$

until the temperature in the system starts to fluctuate about the reference value T_{ref}.

To qualify a gas atom as belonging to a cluster, the total relative energy of the chosen atom interacting within the Verlet sphere about the atom was calculated. If the total relative energy fell into the selected criterion, the atom was counted as belonging to the cluster, and the procedure was moved to the next particle in the Verlet list of the previous particle and all stages were repeated.

The atomic system as evaluated to an equilibrium within several thousand time steps where the time step was selected to be equal to 0.005 a(m/e)1/2, at a given temperature T and then the distribution function was calculated within 104 times steps at each 400 time steps. The final result was averaged over the samplings.

Figure 4.1 shows a phase diagram in coordinates of n–T. Curves 1 and 2 correspond to the coexistence lines vapor-liquid (binodal) of the liquid where atoms interact via a Lennard-Jones potential calculated by the perturbation theory [1] and obtained in experiment [4]. Curve 1 is the spinodal calculated in [1]. The symbols in Figure 4.1 correspond to MD simulations [5]. The calculated results are shown in Figure 4.2 in comparison with the results of a similar calculation where another criterion based on timing was used [5].

Comparing the two criteria shows that the energy criterion used in this chapter significantly reduces the density of clusters compared to those calculated in [5] where the authors used a time analysis. The

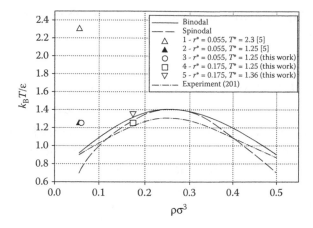

Figure 4.1 Phase diagram in r^*–T^* coordinates. I, I': binodal and spinodal; 2: experiment [4]. Points 1 and 2 [5]; points 3–5, this work. (Data from Vargaftik, N. B., *Tables on the Thermophysical Properties of Liquids and Gases*, 2nd edn, Halsted Press, New York 1975; E. E. Polymeropoulos, J. Brickman, *Chem Phys Lett* 92, 59–63, 1982.)

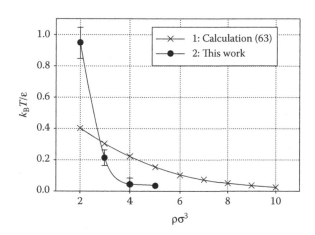

Figure 4.2 Cluster density ($T^* = 1.25$; $\rho^* = 0.55$). (Data from E. E. Polymeropoulos, J. Brickman, *Chem Phys Lett* 92, 59–63, 1982.)

energy data are in good agreement with the results of papers [6] where a similar criterion was used based on the relative complex energy.

It is worth mentioning in conclusion that the existence of clusters slightly reduces the pressure in the system in an undersaturated gas. For the parameters $T^* = 1.25$, $\rho^* = 0.175$, the reduction is about 24%. The pressure calculated by this correction gives the following result: $Ps^* = 0.16$, which is closer to the experimental value

$P_{exp}=0.107$ [4] than that without correction where the calculation gives $P^*=0.218$.

4.3 Kinetics of Cluster Formation in Dense Supersaturated Vapor

Calculations were carried out in microcanonical (NVE) and canonical (NVT) ensembles and in a special ensemble where the temperature was controlled by a thermal bath algorithm.

At the beginning, the system was equilibrated in the single-phase region above the coexistence curve and then suddenly the system was quenched into a region below the spinodal.

Relaxation of the strongly nonequilibrium state obtained by such a quench procedure was studied by obtaining the distribution function of the cluster sizes:

$$n_k = \frac{N_k}{\sum\limits_k N_k},$$

where N_k is the number of clusters of size k (i.e., consisting of k-atoms). Since the relaxation of the system as studied, the averaging was carried out over the ensemble of M initial states ($M=10$–1000). The limitation of the computing resources was limiting the total number of atoms in the basic cell. Figures 4.3 through 4.8 represent the kinetics of the clusters in various ensembles.

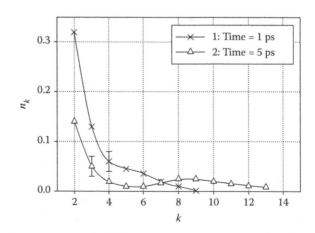

Figure 4.3   Cluster distribution function for a 3D isothermal system.

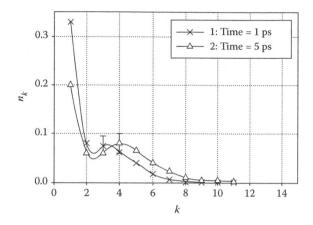

Figure 4.4   Cluster distribution function for a 2D isothermal system: (1) time = 1 ps; (2) time = 5 ps.

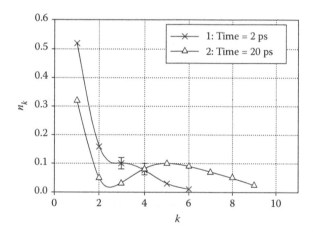

Figure 4.5   Cluster distribution function for a system with thermostat for times of 2 and 20 ps.

The Brownian dynamics method was used for modeling cluster formation kinetics in a vapor in mixture with a dense non-condensing buffer gas [1–3]. The interests in this problem were related to those observed in calculation in NVE ensembles oscillations of the kinetic and potential energies (Figures 4.9 and 4.10).

Figure 4.11 shows the evolution of the total energy in the system for four different friction coefficients that correspond to four buffer gas densities at $T^* = -0.5$ and $\rho^* = 0.15$ (in LJ units). The method of calculation was similar to those in the NVE ensemble.

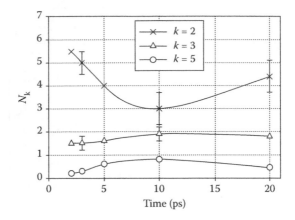

Figure 4.6   Cluster dynamics for an NVE ensemble.

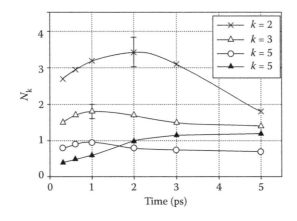

Figure 4.7   Cluster dynamics for an isothermal system.

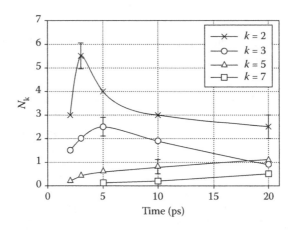

Figure 4.8   Cluster dynamics for a system with thermostat.

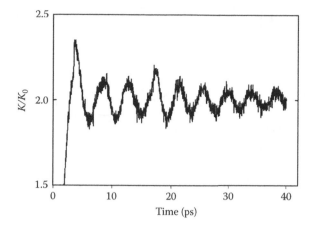

Figure 4.9   Variations of the kinetic energy for a system with $\rho^*=0.32$; K_0 is the initial kinetic energy.

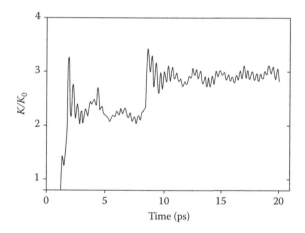

Figure 4.10   Variations of the kinetic energy for a system with $\rho^*=0.32$; K_0 is the initial kinetic energy.

The characteristic feature of the curves in Figure 4.11 is the non-monotonic dependence of the rate of cooling and the "stationary" value of the total energy on the friction coefficient (gas density).

Figure 4.12 represents the gas densities in the system at two time instants: $t/\Delta t=500$ (curve 1) and 3000 (curve 2). From this figure one can see that at the initial time periods the average cluster size is relatively small (<20). However, as with the time goes by the average

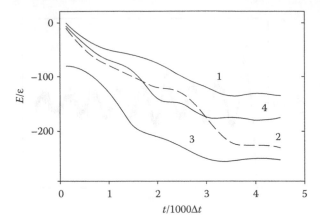

Figure 4.11   Time dependence of the total energy: 1, $\gamma=0.1$; 2, $\gamma=0.3$; 3, $\gamma=1.5$; 4, $\gamma=5.0$.

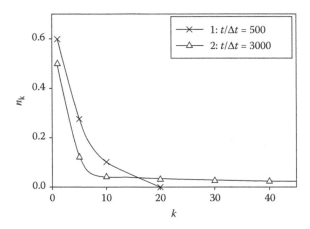

Figure 4.12   Cluster density at $t/\Delta t=500$ (1) and 3000 (2): $\gamma=1.5$.

cluster size quickly grows and the number of small clusters drops since they are "eaten up" by the large clusters. Figure 4.13 shows the dependence of the average cluster size on time for three friction coefficients.

A qualitative comparison with experiments [5] shows that the buffer gas accelerates the rate of cluster formation. In addition, it was experimentally obtained that at an increase of the buffer gas pressure the average cluster size increases at the beginning but then after reaching a certain limit, ceases changing [5]. The latter result also can be observed in Figure 4.11, where the rate of cooling and total energy have a maximum. It comes from this latter fact that the

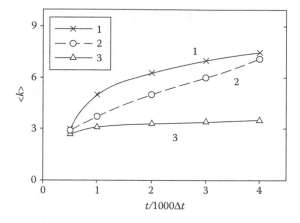

Figure 4.13 Time dependence of the average cluster size: (1) $\gamma=1.5$; (2) $\gamma=0.3$; (3) $\gamma=0.1$.

average cluster size would also have a maximum since cluster formation becomes more impeded at high gas densities.

4.4 Summary of Chapter 4

1. The dynamics of cluster formation in an undersaturated dense gas was calculated by an MD method for the following gas parameters: gas density $\rho\sigma^3=0.055$ and 0.175, and temperatures $k_B T/\varepsilon=1.25$, 1.36, and 2.3 (ε and σ are the Lennard-Jones parameters).

   It was shown that near the coexistence curve of gas–liquid the large clusters with sizes larger than ten were unlikely to exist.

2. The MD method was used to study the kinetics of cluster formation in a dense supersaturated vapor with the following parameters: $\rho\sigma^3=0.32$ and $k_B T/\varepsilon=0.4$. It was shown that for an NVE ensemble, the characteristic was an oscillating character of condensation. A qualitative agreement with spinodal decay theory was obtained.

3. A distribution function on cluster sizes was calculated for a system containing condensing vapor and a noncondensing buffer gas. Qualitative agreement with experiment was obtained for the average cluster size on the density of buffer gas.

References

1. F. F. Abraham, Computational statistical mechanics: Methodology, applications and supercomputing, *Adv Phys* 35, 1–111, 1986.

2. M. P. Allen and D. J. Tildesley, *Computer Simulation of Liquids*, p. 385. Oxford University Press: New York, 1987.

3. D. W. Heermann, *Computer Simulation Methods in Theoretical Physics*, 2nd edn, Springer, 1990.

4. N. B. Vargaftik, *Tables on the Thermophysical Properties of Liquids and Gases*, 2nd edn, Halsted Press, John Wiley: New York, 1975.

5. E. E. Polymeropoulos and J. Brickman, Molecular dynamic study of the formation of argon clusters in the compressed gas, *Chem Phys Lett* 92, 59–63, 1982.

6. V. M. Bedanov et al., Equilibrium vapor pressure over the critical nucleus, *Chem Phys* 6, 997–999, 1987 (in Russian).

CHAPTER 5

Kinetics of Cluster Formation on Surfaces

In this chapter, the interaction of gas atoms and molecules with surfaces is studied using computer simulation. Classical molecular dynamics (MD) was applied for simulation of a fast atom collision with a surface, and the probability of a gas atom to be captured by the wall (sticking coefficient) was calculated. We begin with a broad review of analytical and numerical calculations of the sticking coefficient of gas atoms and clusters to the surfaces, followed by a discussion of the deposition of small metal clusters on surfaces. Collision and surface sputtering by large energetic clusters is discussed next. The second section addresses cluster formation on solid surfaces. The kinetics of the formation of adsorption layers on solid surfaces based on the atomistic theory of nucleation, given in this section, shows that cluster formation is a significant part of thin film growth on solid surfaces.

5.1 Interaction of Gases with Solid Surfaces

Formation and growth of clusters on the solid surfaces plays an exclusive role for all processes of film growth that are deposited from vapor, in a simple liquid or multicomponent fluid [1]. Particle interactions with surfaces are also important for the air and gas dynamics processes of planes and in vacuum technology [2]. Recently, thin film deposition processes from gases containing clusters have been attracting more researchers [3].

As indicated in [4], the surface–gas system is a condensed matter system where many-body interactions play a crucial role. The main theoretical methods are numerical simulation methods such as MD and Monte Carlo (MC). Analytical studies of the processes occurring at the gas–solid surface interface are exceptionally rare (see e.g., [5,6]). On the other side, an experimental technique was developed that can in principle provide a detailed picture of the atomic processes occurring at the boundary between liquid and gases [7–14].

5.1.1   Sticking of atoms and clusters to solid surfaces

The interaction of gas particles with a surface eventually leads to the sticking of the gas particles to the surface, surface diffusion and the formation of surface clusters and growths of them into islands, and thermal desorption and/or evaporation of the adsorbed atoms from the interface to vacuum. Gas particles include atoms, molecules, and clusters of atoms. The sticking coefficient was analytically calculated in [5,6].

Classical and quantum theories of interaction of a single gas atom with a solid surface were given by Goodman and Wachman in the first fundamental book in the field [6]. The authors have made several important conclusions: (1) the gas–surface collision should be defined as a process where a gas particle hits the surface and changes its velocity component normal to the surface to a negative sign, and this collision should occur between two consecutive gas–gas collisions; (2) the sticking of a gas particle (an atom or a molecule) should be corrected to the processes where the gas particle hops several times after the first collision with the surface, until it is stuck at a position that can be far from the first strike position; (3) the atomic motion of the surface atoms should be taken into account because, depending on the masses of incidence and surface atoms, the surface atoms can be "struck" several times before the gas atom changes its velocity component sign to opposite and then "struck" again several times after the velocity component changes its sign.

There will be also another part of the gas particle that strikes the surface atoms several times and returns to gas, without being stuck to the surface. The third class of gas atoms experience several collisions and exchange kinetic energy with the surface atoms but do not stick for a time that is beyond the available computation time.

The gas–surface collisions change the properties of both the gas and the surface. The gas molecules will have changes in the velocity distribution function and the surface can be subject to modification.

Most of the information about the interaction of gas with surfaces is obtained by experiments with molecular beams (including atomic ones) that are torques exerted to the targets suspended on torsion balances or forces acting on the microbalances.

The dynamics of scattering the gas directed toward the surface is mainly dependent on the interaction energy of an individual gas atom with the surface atoms, and is less dependent on the interaction of gas—gas atoms among themselves. The motion of the surface atoms can be represented in a harmonic approximation, by the modal properties of the solid that are defined by the interactions of the solid atoms among themselves.

Usually, the interaction of gas molecules with the surface is approximated by some appropriate model potential, which can be a simple Lennard-Jones (LJ) type or a Morse type, which is discussed in detail later in this chapter.

The incident energies of gas atoms in collision experiments with molecular beams are in the range $0 < E_i < 40\,eV$, which are so small that they do not normally cause sputtering and erosion of the surfaces.

Therefore, in principle, the parameters of the gas atom interaction potential with the surface and the parameters of the modal structure of the solid can be reconstructed from experiments with the molecular beam scattering from the surfaces, or they also can be obtained by using the experimental data of the sputtering and surface erosion data by gas atom bombardment.*

The geometry and the properties of molecular beams are as follows: the beams are assumed nondiverging, that is, having no angular divergence, and having a Maxwellian velocity distribution function that is related either to an expanding jet or an effusive Maxwellian stream.

The experiments should be capable of obtaining (1) the velocity distribution of the incident beam and (2) the velocity distribution of the scattered beam.

The modern classical theory of gas–surface scattering at low energies was developed by Cabrera and Zwanzig for a one-dimensional lattice model and by Goodman, Chambers, and Kinzer for three-dimensional lattices (see e.g., [6] and references therein). The latter model was still restricted to a one-dimensional motion of the gas atoms and to zero initial temperature [6].

The first fully three-dimensional Monte Carlo trajectory simulations were conducted by Oman and his co-workers [39,40], and the results of such simulations were very important for the development of the gas–surface scattering theory. Although these first MC

---

* Jim Belak's private communication.

simulations were too time consuming and cumbersome, two types of scattering mechanisms of a gas molecule from a solid surface were proposed based on MC simulation data: a thermal and a structure scattering type, and a classical rainbow scattering mechanism [6].

One-dimensional results were combined with a simple flat surface assumption that was believed to be applicable to the thermal type of scattering; first a "hard-cube" model of the gas–surface scattering was developed [6]. A more realistic "soft-cube" model was then introduced that replaced the hard sphere interaction to a more realistic soft variation of the potential energy near the surface, for the direction normal to the surface, and implied a nonzero surface temperature and a characteristic surface vibration temperature, in analogy with the Debye temperature.

The main progress in this direction was obtained with fundamental justification of the classical hard- and soft-cube models by a quantum mechanical theory. The classical rainbow theory was represented as an envelope of many quantum mechanical diffraction peaks.

Also, the cube models do not provide rigorous descriptions of the gas–surface interaction; they are still the main source of correlation of a large set of experimental data with simple expressions, assuming that the mechanism is close to the thermal type of collisions.

In any analytical theory or a computational experiment, three parts of the gas–surface collision process need to be clarified: (1) a gas, (2) a solid target, and (3) the interaction between these two parts. The gas particles were treated as structureless point particles, thus discarding all electronic degrees of freedom in the atom.

The surface usually is modeled by a cube lattice model, with a nearest-neighbor harmonic spring connection, or by a flat surface model. For example, in a soft-cube model the surface is represented by a single mass (cube) that is attached to a rigid wall via a harmonic spring.

In more elaborated models, the surface is modeled as a collection of point mass connected by springs located at positions of an ideal fcc or bcc crystalline, thus describing a realistic crystalline structure. However, the bulk structure in not taken into account. Surface roughness is also not taken into account. Therefore, a polycrystalline surface that consists of many disoriented grains would look like an ideal defectless surface without realistic surface effects. No adsorbed atoms are usually included in the account.

Sticking or trapping or absorbing a gas molecule to a surface is considered to be a collision where the total energy of the molecule, including the internal kinetic energies of translational, rotational, vibrational, and potential energies of interaction with the surface, becomes negative. The zero level of energy is chosen at the farthest distance of the molecule from the surface [6].

In [15–18], a collision of a gas particle with an idealized model of a surface was studied by analytical theory. The gas particles were regarded as "stuck" to the surface after the collision if a reflected particle had energy $E \leq E_c$, where E_c is the critical energy. All collided atoms with the kinetic energies larger than E_c were considered as reflected. Assuming the incoming flux of molecules has a Maxwellian velocity distribution function and introducing a sticking coefficient as a ratio of the number of stuck particles to the total number of collided to the surface, the authors derived the following expression:

$$\alpha = 1 - \left(\frac{E_c^2}{2R^2T^2} + \frac{E_c}{RT} + 1 \right) \exp\left(-\frac{E_c}{RT} \right), \tag{5.1}$$

where:

R is the universal gas constant

T is the absolute temperature and it was assumed that $E_c = E_d$, where E_d is the depth of interatomic potential for interaction of the gas particles with the surface.

The above model was elaborated on further in [17]. The surface was represented as a system of independent oscillators having a Boltzmann distribution of energies. The coefficient of sticking was obtained as the following expression:

$$\alpha = \frac{1}{2} \left\{ \left(1 + erf \left[\frac{1+\mu}{2\sqrt{\mu}} \left(\frac{E_d}{k_B T} \right)^{1/2} \left(1 - \frac{1-\mu}{1+\mu} \left(1 + \frac{E_i \cos^2 \theta_i}{E_d} \right) \right)^{1/2} \right] \right) \right\}, \tag{5.2}$$

where:

$\mu = m_g/m_s$, m_g, m_s are the atomic masses for gas and surface

E_i is the energy of incoming particles

T is the surface temperature

This expression is valid for $\mu \ll 1$. Additionally, the authors in [17] showed that the above formula agrees well with their own experiments at low incoming angles. It should be mentioned that, according to [15], experimental measurements of the sticking coefficient are complicated at large incoming angles.

In [5,18], the scattering process of gas atoms from a surface was considered as a one-dimensional stochastic process in which the gas particle releases its energy in a series of consecutive collisions. In that case, the kinetic equation transforms into a Fokker-Plank equation and its solution can be used to calculate the sticking coefficient as follows:

$$\alpha = \frac{1}{\left[\dfrac{T_g}{\mu^*} + \dfrac{1}{2} \left(1 + 4 \dfrac{T_s}{\mu^*} \right)^{1/2} \right]},$$
(5.3)

where:

T_g, T_s are the gas and surface temperatures
μ^* is the energy loss of a particle in single collision

Interactions of gas particles with *realistic* models for the solid surfaces were studied by theory and simulation in the papers [18–31].

A Monte Carlo method was used to calculate the sticking coefficient of helium atoms to the surface of solid argon and nitrogen at cryogenic temperatures by Leuthäusser and Kryukov [18,19]. The target surface was modeled by a three-dimensional crystalline lattice at zero temperature and the binary collisions of the gas atoms with the lattice were assumed. Surface roughness was taken into account which was introduced by multiple reflections of a gas particle within a single act of collision. As expected, the sticking coefficient increased with the increase in the number of reflections from a rough surface.

However, Zhuk [20] found that the dependence of the sticking and accommodation coefficients on the number of multiple reflections from the rough surface is rather weak and should not increase the sticking probability.

Adelman and Doll [19–24] proposed a new method of stochastic trajectories in which the interactions of the gas particle with the surface and the mutual interactions of surface atoms are taken into account using realistic many-body interactions. For doing that, the authors separate the surface area into a series of m zones that are dependent on the distance to the collision central zone, where the interactions are fully included in a model pair-additive approximation; interactions of the surface atoms located outside of the central zone are taken into account via a Langevin approximation where the effect of the local interactions is included using a friction force and a random force with a Gaussian shape. Such a method was obtained to be very useful after it was further developed in a series of papers by Tully and coworkers in [24–31]. In [24,25], interactions of He and Ar atoms with a tungsten (W) surface were simulated by the method where the W surface was replaced by four atoms of the central zone.

For the scattering of gas Ar_n ($n = 5$–26) clusters with a Pt(111) surface the authors [26] used 72 atoms in the central collisional zone. Incident energies of 0.1–0.5 eV per cluster atom were used. This study showed that the cluster fragmented almost fully during the impact with the surface. The angular distribution of the scattered atoms for the larger cluster was obtained to be close to tangential and almost independent of the incident angle. The kinetic energy of the tangential scattered atoms is the largest, and decreases while the angle approaches the normal to the surface. Such behavior was explained by multiple collisions inside the cluster as well as between the cluster atoms and the surface.

To study the interactions of NO molecules with the surfaces of Ag(111), Pt(111), and LiF, the authors of [27–29] used 32 atoms in the central collision zone.

Some details of the calculations [24–30] will be brought up in the following. The trajectories of the atoms obtained numerically were the initial gas atoms that were bombarding the surface at a certain incoming angle. A Lennard-Jones (LJ) (12-6) interaction potential was used to take into account the interactions of the gas particles with the surface. The sticking coefficients calculated by different authors who used different approaches are given in Table 5.1.

However, since the above method uses some sort of series over small deviations of atomic positions, it is naturally not applicable to collision events where the atomic positions are significantly disturbed in the central zone. This restriction did not allow studying the highly energetic processes where the gas atoms were able to penetrate into the surface.

A new method of stochastic trajectories was developed in [26] and applied to the study of scattering of small Ar gas clusters with

Table 5.1   Sticking Coefficients Calculated by Different Methods

Type of Gas Atoms or Clusters	Type of Surface	T_g, K	T_s, K	α, Sticking Coefficient	Calculation Method[a]	References
He	Ar	77	0	0.90	1	[19]
		300	0	0.60		
	N$_2$	77	0			
		300	0			
Ar	Pt	1100	300	0.14	2	[26]
Ar$_2$				0.37		
Ar$_3$				0.41		
Ar$_6$				0.40		
Xe	Pt	840		0.86	2	[29]
		2160	95	0.37		
		4440		0.12		
		—		0.04		
K		5000	1150	1.00	2	[17]
Na	W	5000	1226	1.00		
K	W	4000	1200	0.50	3	[20]
		8000		0.18		

[a] 1, Monte Carlo method; 2, stochastic trajectory method; 3, direct numerical calculation.

thermal energies from a Pt(111) surface. The cluster impacts have resulted in the breakup of clusters and sticking of some of the cluster atoms to the surface, as well as to a partial reflection of cluster atoms from the surface.

The sticking coefficient of the cluster impact was calculated to be in the range of 0.14–0.41, depending on the cluster size. The portion of the stuck atoms grows with the cluster size, which was explained by the increase in the number of intracluster collisions.

Comparison of the angular distribution of the number of atoms scattered from the surface with the experiments [26] on a Fe(100) surface in which $(N_2)_n$ $(n = 4800)$ was not satisfactory.

The conventional theory of the gas–surface interaction was based on hard-cube and soft-cube models.

5.1.2   Cluster ion deposition of thin films

Starting from the pioneering work of Takagi and his coworkers at Kyoto University in 1972 [3,32–38], an ionized cluster beam technique has received considerable interest and critical attention as a new and prospective method for fabricating high-quality thin films by depositing accelerated cluster ions containing ~1000 atoms. Figure 5.1 shows a schematic diagram of the cluster ion source. It seems that this method allowed one to fabricate metal, semiconductor, insulating, and magnetic thin films on solid cold substrates. The material for thin film fabrication was placed in a crucible and then evaporated by heating with a filament of electron beam into a vacuum chamber and then adiabatically expanded through a small aperture diaphragm. The new type of ion source was named the *vaporized-metal cluster ion source* and showed a high deposition rate for thin films. Clusters (aggregates) of metal atoms were generated inside of a heated crucible and extracted via a small aperture into vacuum by adiabatic expansion. After the vapor beam passed through the small throat, it was ionized by the electron irradiation and then accelerated toward the target that was kept at much lower potential.

Among the advantages of the new source, the authors named small charge-space effects, small charge-mass ratio, increased surface diffusivity, rise of the local temperature, and high sputtering rate that might lead to better quality thin films. The formation of clusters in the gas was predicted to be relevant for some of the metallic materials, such as Pb, where the authors predicted the

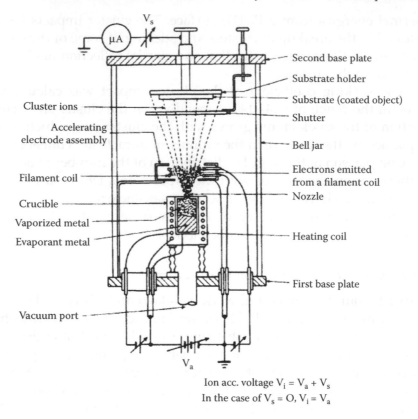

Ion acc. voltage $V_i = V_a + V_s$
In the case of $V_s = O$, $V_i = V_a$

Figure 5.1   Schematic representation of the vaporized cluster ion source. (From T. Takagi, et al., *Proceedings of the 2nd International Conference on Ion Sources*, p. 790, Österreichishe Studiengessellschaft für Atomenergie, Vienna, 1972; T. Takagi, et al., *Japan J Appl Phys Suppl* 2(1), 427–430, 1974; T. Takagi, et al., *Thin Solid Films* 39, 207–217, 1976; T. Takagi, et al., *Thin Solid Films* 45, 569–576, 1977; T. Takagi, et al., *Thin Solid Films* 92, 1–17, 1982; T. Takagi, et al., *App Optics* 24, 879–882, 1985; T. Takagi, et al., *Pure Appl Chem* 60(5), 781–794, 1988.)

formation of clusters with an average size of 8×10^2 at the conditions of $P_1 = 1$ Torr and $P_2 = 10^{-5}$ Torr.

According to the Classical Nucleation Theory (CNT), such vapor can easily be overcooled, so that the vapor temperature becomes much lower than the vapor dew point. This would instantly lead to the formation of a new phase. The nuclei of the new phase are represented by microscopic small clusters (aggregates) containing anywhere from several hundred to several thousand atoms depending on the level of overcooling.

The clusters were ionized by an electron impact and accelerated in a high electric field and forwarded to a target substrate for film formation. This method was not well suited to the formation of metal films due to the high surface tension of the metal clusters, which impeded the formation of large clusters. Therefore, as it was confirmed later [38], the concentration of large metal clusters was negligibly small in the beam. However, even a low number of small and intermediate metal clusters were sufficient to deliver a significant amount of kinetic energy to the surface and to adsorbed atoms so that they were capable of significantly improving the overall quality of the thin films fabricated by such method. This method was significantly improved by Haberland and coworkers [39–42], who developed a new gas discharge ion source that could have easily created large metal clusters in a carrier gas by a magnetron sputtering. In such a source, almost 80% of the metal atoms were ionized within the condensation zone, which makes an additional ionization unnecessary.

Figure 5.1 shows a schematic of the original Takagi's equipment [13]. A cluster ion containing about 1000 atoms, accelerated to several keV, collides with the solid surface. The cluster melts during the collision due to the kinetic energy, and the cluster atoms easily can migrate on the surface. Microcracks and microholes on the surface will be easily filled in by the adsorbent atoms and, therefore, the adhesion properties of the thin films will be much higher than those deposited from an atomic vapor.

Figure 5.2 shows a simplified schematic of the physical processes occurring during an energetic cluster ion (ECI) deposition that starts with descending (lowering) of a metal cluster containing from several tens to several hundreds of metal atoms, with kinetic energy of 0.1–10 eV per atom. At a lowest cluster energy of 0.1 eV/atom, the cluster deposition process can be characterized as a soft landing where the cluster survives the impact, with a small or no change to shape, crystallinity, and the number of atoms. At a higher kinetic energy of 1–10 eV/atom, the cluster atoms acquire kinetic energies higher than the binding energy, with a typical values of 5–7 eV/atom for most of the materials.

Therefore, an impact of such cluster on a surface can completely destroy the cluster and trigger an intense mixing of its atoms with those belonging to the substrate. The mixing of cluster atoms with the substrate would make a much stronger adhesion of the film that will be formed by the deposited cluster atoms. Therefore, one of the

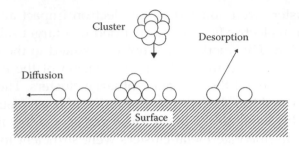

Figure 5.2   Schematic of the processes occurring at the surface during the energetic cluster ion deposition. (From H. Haberland, et al., Thin-film growth by energetic cluster impact (ECI)—Comparison between experiment and molecular-dynamics simulations, *Mater Sci Eng B* 19, 31–36, 1993.)

advantages of the ECI method was the formation of strong adsorption layers, with less porosity and higher density and surface flatness and, as a result, better electric characteristics.

Figure 5.3 shows a schematic diagram of experimental equipment developed by Haberland and coworkers at the University of Freiburg-im-Breissgau in Germany [39]. Part of the deposited atoms will be separated from the rest of the adsorbed atoms (adatoms) and will be moved to large distances via surface diffusion and can also be partially desorbed into vacuum. Finally, the adsorbed atoms can easily create epitaxial layers repeating the symmetry of the substrate lattice. A negative potential is supplied to two electrodes C1 and C2, which generate a glow discharge between them at a gas pressure of 0.1 Pa. A large potential drop is located near the negative electrode (cathode potential drop) where positive ions are accelerated toward the cathode and bombard the surfaces causing erosion.

The sputtered surface atoms are carried into the carrier gas (mostly Ar) where collisions with the gas slow down the sputtered atoms. The carrier gas transports the sputtered atoms to a condensation area that can either be cooled down to a low temperature (e.g., be at a liquid nitrogen) or heated up to a higher temperature. Two apertures A1 and A2 control the gas flow in and out of the condensation area.

After the flow passes the chambers with differential pumping, the central part of the gas flow is cut by a skimmer A3 and accelerated toward the target. The intensity of the deposition is measured

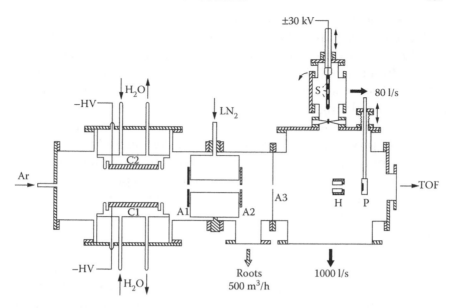

Figure 5.3 New experimental equipment developed by Haberland's group at the University of Freiburg-im-Breissgau for the energetic cluster ion (ECI) deposition process. (From H. Haberland, et al., Thin-film growth by energetic cluster impact (ECI)—Comparison between experiment and molecular-dynamics simulations, *Mater Sci Eng B* 19, 31–36, 1993; H. Haberland, et al., Molecular-dynamics simulation of thin-film formation by energetic cluster impact (ECI), *Z Phys D* 26, 229–231, 1993; H. Haberland, et al., Thin-films from energetic cluster-impact—Experiment and molecular-dynamics simulations, *Nucl Instrum Methods Phys Res B* 80(1), 1320–1323, 1993; H. Haberland, et al., *Phys Rev B* 51, 11061–11067, 1995.)

by a movable lever R. The cluster beam can also be analyzed by a time-of-flight (TOF) mass spectrometer which is not shown in the figure.

The size of the clusters deposited onto a film is analyzed by electron microscopy. Three target samples can be located on the lever P. The sample is placed in front of the heater H. The target can be biased with a DC potential of ±6 kV. The equipment also allows one to ground the target and supply a high voltage to two cathode electrodes. This method can generate a cluster beam of Al, Cu, and Mo containing clusters with an average size of ~10^4 atoms.

For sputter insulating materials, a radio-frequency discharge should be used additionally.

Theoretical studies of cluster ion beam deposition were provided by MD computer simulations [43–47] and Monte Carlo [48–50].

Müller [43] developed a two-dimensional MD model using a two-dimensional Lennard-Jones interaction potential acting between both the cluster and substrate atoms and applied his model to simulating a film deposition onto a layer of its own atoms. Although this two-dimensional model was an oversimplification of the real processes, the author clearly differentiated two distinct mechanisms of deposition. At low cluster translational energies of $E/N < 0.1 \, \varepsilon$, where ε is the depth of the interaction potential between two atoms, the growing film had an amorphous "spongy" structure containing multiple pores. At $E/N > 1.5 \, \varepsilon$, the film grows and has a perfect epitaxial structure.

In papers [44,45], deposition of Si clusters onto a Si(111) was simulated by MD at low cluster kinetic energies <1 eV/atom. Si atoms were interacting via a Biswas–Hamann potential where two- and three-body terms could have been taken into account separately. The three-body term was enhanced by a factor of 2.5 in order to make it applicable to an amorphous phase of Si. As a result, the authors obtained that a metal phase of Si with six and more nearest neighbors become less energetically favorable. The remaining six parameters of the interaction potential were chosen from the best description of volume, surface, and Si defect properties.

Clusters contained 1, 8, 33, and 50 Si atoms. The substrate was modeled by 768 and 192 mobile atoms, and 384 atoms were fixed at the lattice position and were immobile. The main results of these papers were the calculation of the mobility of the deposited islands on the surface temperature and cluster kinetic energy. The cluster kinetic energy was addressed as the most influential parameter affecting the mobility.

Collisions of metal clusters with surfaces were simulated in [46]. Clusters of Cu, Ni, and Al having a maximum energy of ~80 eV/atom and sizes from 4 to 92 atoms were forwarded toward the surface built of about ~4000 own atoms. The Interaction potential was combined by a Moliere potential [47] at short distances and an EAM type at average distances. At small translational cluster energies of the order of 1 eV/atom, the collision leads to the formation of a hemispherical island on the top of the surface. At energies of 25 eV/atom, the surface experienced drastic radiation damage. If the cluster atoms were lighter than those of the lattice, then most of the cluster atoms would be reflected back to vacuum.

Some papers deal with the development of a new time-dependent Monte Carlo method that permits getting beyond a typical binary collision approximation (BCA) [48–50]. To achieve this goal, the atoms confined within a central collision zone limited by three cut-off radii from the center for which the applicability of BCA was not certain were studied by MD.

The main result of such approach consists in a qualitative description of the sputtering of amorphous carbon by argon clusters Ar_n ($n = 10 \div 200$), the depth of penetration of an Ag_n ($n = 190 \div 500$) to the same substrate, and aluminum cluster Al_n ($n = 100, 500$) deposition on a Si(111) surface. The atoms of the target were placed randomly in accordance with the average material density. The same interatomic potential was used for all the studied materials where the depth of the potential was adjusted to a specific material. However, such approximations do not seem to be consistent since the density in the central collisional zone can change dramatically within the impact of a heavy and large cluster, which makes the above model [48–50] less attractive.

5.1.3 Collisions of energetic clusters with targets

The theory of sputtering of materials at interactions of high-energy ions with a target was developed by Sigmund and coworkers [51,52], Thompson [53], and reviewed in another paper [54]. A theory of sputtering by a cluster ion does not exist today, unfortunately. Although many features of such collisions such as significant release of the kinetic energy of collision, heat, and pressure are the same as those of a single ion collision, quantitative characteristics such as the rate of surface erosion per cluster atom, Y, would not comply with the Sigmund's linear cascade theory.

In [51], a further development of the linear theory for molecular ions was given that proposes a linear dependence of the energy release on the number n of atoms in a molecular (cluster) ion: $E \sim n \times E_1$, where E_1 is the energy released by a single ion collision.

Thompson [53] studied the erosion of Ag, Au, and Pt with high-energy molecular ions P_n, As_n, Sb_n, Bi_n ($n = 1, 2, 3$), within a molecular ion kinetic (translational) energy interval of $10 \div 250$ keV.

Comparison between the experimental and calculated data showed that the results at low ion energies can be explained satisfactorily by the linear cascade theory of a single ion collision.

After the cluster energy achieves a certain threshold, a new thermal evaporation mechanism (the so-called thermal spike mechanism) contributes to surface erosion and therefore the rate of erosion (Y) changes its slope from a linear cascade to a much steeper one. One more mechanism the author indicated was the evaporation of large droplets or clusters from the target's material.

A similar conclusion was obtained by the authors of a review [54] who analyzed the irradiative erosion of a carbon target by water clusters consisting of 25 water molecules that were bombarding the surface with a velocity of ~10 km/s.

The sputtering yield (Y) of a Au target bombarded with accelerated cluster ions of Ar_n $(n=300)$, with energy of 28.8 keV, was experimentally measured to be $Y = 105$ (atoms/ion) [55]. The authors discussed new applications of the cluster ion technique such as ion implantation, improvement of surface quality, and surface treatment by cluster ions.

The ejected flux from the irradiated material contains neutral atoms, singly charged ions, and some amount as small clusters of the target material. The presence of clusters in the ejected flux was experimentally proved in time-of-flight (TOF) spectroscopy with energy resolution studies in [56–58]. Polycrystalline Ag was bombarded by 5 keV ions [56]. The clusters ejected from the surface were ionized by an excimer laser based on Ar F ($hv = 6.4$ eV). The kinetic energy of the particles to be detected was obtained from the time-of-flight of particles between the sputtering ion pulse and the ionizing laser pulses. The relative sputtering yield was obtained as a dependence on the cluster size. The kinetic energy distribution of the ejected flux was fitted by an $E^{-1.8}$ for Ag single atoms (monomers) and by an $E^{-2.3}$ for dimers and other cluster sizes.

The target was a polycrystalline Al and the following irradiation Ar^+ ion energies: 8 keV in [57] and 3.9 keV in [58]. The sputtered flux consisted of monomers, dimers, and trimers of Al [57]. The energy dependence of the ejected atoms was fitted by an E^{-2} power law, whereas dimers and larger clusters were well fitted by an E^{-3} curve. Various energy dependences E^{-n} of the cluster sizes $n = 2 \div 6$ were observed in [57,58] and the results for various power exponents n are collected in Table 5.2.

Table 5.2   Energy Dependence of Relative Abundance of Small Metal Clusters

Type of Cluster	Power Exponent, n
Al	2.6
Al_2	3.3
Al_3	4.4
Al_4	2.8
Al_5	4.1
Al_6	3

Source: A. Wucher, et al., *Nucl Instrum Methods Phys Res B* 82, 337–346, 1993.

5.2   Cluster Formation on Solid Surfaces

Thin films are attracting much attention in view of their important applications in microelectronics, as protective coatings against corrosion, and in the fabrication of optically transparent layers. Besides these practical applications, thin films are interesting materials for fundamental studies since they represent an intermediate phase with the physical properties that are characteristic for both two- and three-dimensional systems.

There is also strong interest in a new field of cluster formation and growth of thin films due to the applicability of the theory of nucleation and condensation to this field and an opportunity for much easier experimental observation of thin film growth. Therefore, this field can be an ideal physical model for verifying or rejecting the condensation theories including Classical Nucleation Theory (CNT), mainly due to the ease of observation of the film nucleation, coagulation, and growth by scanning tunneling microscopy (STM) or by a high-resolution transmission electron microscope (HRTEM) that can be installed directly in situ. Atomic resolution imaging of islands on the surface by HRTEM was obtained in [59]. Surface thin film formation occurs via intermediate structures such as islands that move, diffuse, and grow and finally become part of the growing film. Direct observations were of clusters with sizes from 1 to 10 nm on various supports including zeolites, cordierite, and amorphous carbon. Small clusters of Au, Pt, Rh, and Pb were recorded and the following dynamic types were revealed: the smallest clusters were able to rearrange quickly, and "clouds" of adsorbed atoms were involved in structural rearrangement, hopping, and desorption.

Theoretical papers can be classified into the following groups: a) classical theories [8]; b) atomistic theories [60–62,1]; and c) using mean-field theories that can be expressed as kinetic rate equations for clusters of different sizes [63]. The first two groups develop models that predict the concentrations of critical and subcritical nuclei. The nucleation rate can then be calculated as a product of the critical nuclei concentration to a frequency of transition of the critical to the over-critical nucleus.

The critical radius is not used in the mean-field theories that are the basis of the derivation of the kinetic equations. Therefore, to solve the microkinetic equations, one does not need the concept of a critical cluster size. If the kinetic coefficients entered into the equations are known, the solution of the kinetic equations easily gives the distribution function on the cluster size, as well as a microscopic nucleation rate constant.

Numerical investigation methods in the kinetics and thermodynamics of thin film nucleation and growth should be able to predict the structure, dynamics, rate constants, and evaporation rates of the surface particles as small clusters by the Monte Carlo and stochastic Langevin equations [54,64].

Since the Classical Nucleation Theory (CNT) was discussed in Section 5.2.1 and is well documented, we start with the atomistic nucleation theory that was developed in the works of Rhodin and Walton [65] and Hirth and Pound [8].

5.2.1   Atomistic theory of nucleation

Walton et al. [65] demonstrated that the concentration of clusters of atoms (molecules) can be derived from

$$\frac{n_g}{n_0} = \left(\frac{n_1}{n_0}\right)^g \exp\left(\frac{E_g}{k_B T}\right), \tag{5.4}$$

where:

n_0 is the concentration of the absorption centers

E_g is the energy gain at formation of a g-cluster from g-monomers

Derivation of Equation 5.4 does not take into account various configurations of the g-mer, including isomers. However, it was shown that taking into account such configurations does not make drastic changes into the picture of growth, therefore we will not discuss this issue here. A much more severe shortcoming of Equation 5.4 is the omission of rotational and translational degrees of freedom of clusters, which was corrected by Hirsh and Pound [8].

In a dynamic equilibrium of clusters between themselves, the following equation should be satisfied:

$$\Gamma_g^+ n_g = \Gamma_{g+1}^- n_{g+1},$$ (5.5)

where $\Gamma_g^\pm$ is the probability of capturing of a monomer (for + sign) by a g-atomic cluster, and of losing a monomer (for the $-$ sign) by the g-atomic cluster. Direct collisions of a monomer with a g-cluster contribute to the capture and surface and volume diffusion. Usually, I^+ is defined via a surface diffusion as follows [66–69]:

$$\Gamma_g^+ = \alpha A_g n_1 a \nu \exp\left(\frac{-\Delta G_{\text{diff}}^*}{k_B T} \right),$$ (5.6)

where:

A_g is the area of capture of a monomer by a g-mer
α is the jump length
n is the frequency of the jumps
ΔG_{diff}^* is the free energy of activation of the desorption

A critical nucleus g^* was defined by Walton et al. [65] as a "cluster with the probability of growth less than or equal ½ and which increases the growth probability to higher than ½ after capturing a single atom."

The nucleation rate, according to Walton's theory, can be defined as follows:

$$J_0 = \frac{\alpha a \nu A_{g^*}}{n_0^{g^*-1}} \left(\frac{\beta}{\nu'} \right)^{g^*+1} \exp\left[\frac{(g^*+1)\Delta G_{\text{des}}^* + E_{g^*} - \Delta G_{\text{diff}}^*}{k_B T} \right].$$ (5.7)

Although Equation 5.7 does not include any microscopic parameters of the material, calculation of E_{g^*} is still a very complicated task.

Comparison of the statistical theory with experiment for the nucleation rate for silver on a NaCl surface was made in [70]. The pre-exponential parameter in Equation 5.7 was estimated in the following way. A graphical dependence of $\ln (J_0)$ on $1/T$ was drawn which, according to Equation 5.7, should be a straight line with the slope $Q/k_B = [(g^*+1)\Delta G^*_{des} + E_{g^*} - \Delta G^*_{diff}]/k_B$ that crosses the axis line at $\ln[av A_{g^*} n_0^{1-g^*}(\beta/v')^{g^*+1}]\lim_{x\to\infty}$. The activation energy was estimated for T above 250°C and $Q=3.8$ eV; for lower temperatures $Q=0.7$ eV. Such a jumping behavior of Q was assigned to a strong dependence on the critical size g^* at the transition temperature. Assuming $G^*=1$ at $T<250°C$ and $g^*=3$ for $T>250°C$, we can derive the pre-exponential factor in Equation 5.7: $3\times10^{-27}\,cm^{-1}s^{-1}$ [71] for the following parameters: $n_0=2.6\times10^{15}\,cm^{-2}$, $n=n'=1$ $e12$ (1/s), $A_1=a=n0^1/2$ and for $b=6\times1013$ cm^{-2}s^{-1}. The nucleation rate constant estimated by this model drastically deviates from the experimental value $3\times10^{28}\,cm^{-2}s^{-1}$ that was obtained in [71]. It should be mentioned that the sticking coefficient of a single atom to a g-cluster surface was set to be unity.

5.2.2   Kinetic equations for thin film nucleation

Cluster formation and growth are important processes for fabricating thin films. The processes involving clusters include physical and chemical mechanisms of adsorption, thermal desorption into vacuum, surface diffusion of adsorbed atoms (adatoms) and small clusters. There are three distinct mechanisms (or modes) of thin film growth: (1) an island growth mode or a Volmer–Weber model; (2) a layer-by-layer growth mode (or a Frank–van der Merwe model) that is typical for atoms with strong bonds to the surface; and (3) the layer plus island (a Stranski–Krastanov model)—as intermediate.

The rate of the surface elemental processes shown in Figure 5.4 is limited by an energy barrier. Therefore, the probabilities of them being activated thermally can be defined by an Arrhenius law. For example, thermal re-evaporation occurs at a characteristic time defined by the energy of absorption E_a, $\tau_a \sim v \exp(-\beta E_a)$, where v is a characteristic frequency of surface atomic vibrations for the specific event type, and $\beta=1/k_B T_s$, where T_s is the surface temperature. Various elemental processes have different activation and binding energies indicated in Figure 5.4, where E_d is the migration energy for surface diffusion. Processes involving small cluster formations are characterized by E_j, which is the binding energies of a cluster containing j atoms; where E_i is the binding energy of a "critical" cluster size with i atoms.

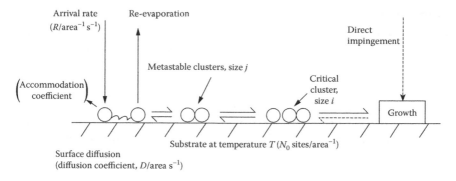

Figure 5.4 Schematic representation of the physical processes involving thin film formation. (From J. A. Venables, *Philos Mag* 27(3), 697–738, 1973.)

Two experimental parameters R (deposition rate) and T_s control the kinetics of the processes, assuming that the energies of the thermal activated processes (E_d, E_a, E_i) and binding energies are known.

Since a real surface can provide many imperfections such as contamination, kinks, ledges, and surface dislocations, these nonideality centers may act as nucleation centers of a nucleation bypassing the usual condensation process via a critical cluster size. These processes are similar to the heterogeneous condensation in the three-dimensional case.

In a series of papers [63,65], Venables et al. have developed a simplified kinetic approach for thin film nucleation and growth for deposition from vapor. The following assumptions were made: (1) The supersaturation ratio of the vapor was so large that the critical sizes of the stable nuclei were between one and six atoms. (2) The dissociation (decay) processes of clusters were neglected, which is applicable to strongly bound clusters. (3) Dimers and larger clusters were immobile. The only moving species were single adsorbed atoms. (4) The surface is homogeneous and isotropic without traps and dislocations.

The model contains three parameters that should be obtained from experiment: D_1 (cm^2/s)—the surface diffusion coefficient, E_A (eV)—the adsorption energy of the atom on the surface, and E_B (eV)—the binding energy of atoms in the cluster. The last two parameters are independent of the cluster size and location and the shape of the cluster on the surface. Although the actual adsorption energy cannot be a constant, since electron microscopy often reveals bright spots on the surfaces, including kinks, edges, and dislocations, there are still wide areas (terraces) that are homogeneous which can be considered as having uniform coating where constant E_A is a good approximation. Fortunately, these three parameters were available from experiments for a large number of metals [63, Part B]. This energy defines the potential barrier for a single adsorbed atom (adatom) to be desorbed from surface into vacuum at temperature T: $\tau_A = 10^{-13} \exp(E_A/k_B T)$.

According to this model, the concentrations of clusters of different sizes can be obtained by solving the following set of kinetic equations [63]:

$$\frac{dn_j}{dt} = U_{j-1} - U_j \quad (i \geq j > 2), \tag{5.8a}$$

$$\frac{dn_j}{dt} = U_{j-1} - U_j \quad (i \geq j > 2), \tag{5.8b}$$

$$\frac{dn_x}{dt} = U_i, \tag{5.8c}$$

$$U_x = \sum_{j>i} U_j, \tag{5.8d}$$

where:
  R is the condensation rate ($cm^{-2} s^{-1}$)
  τ_a is the re-evaporation time
  n_1 is the surface density per cm^2 of single atoms adsorbed on the surface (adatoms),
  n_x is the total number of clusters of super-critical sizes
  U_j is the growth rate (flux) of a j-mer to a $j + 1$-mer

The adatoms diffuse with the coefficient of D_1 (cm^2/s) at a surface temperature T. Small clusters therefore form a chain of metastable clusters that form by associating single atoms and dissociate by the evaporation of single atoms into the adsorbed layer. At a certain size $i \geq j \geq 1$, the rate of association becomes larger than the dissociation rate and these clusters are called critical. The assumption is that the critical size i^* does not depend on time. U_x is the rate at which a single atom joins a critical cluster and makes it a super-critical one. All clusters of sizes $\geq i$ are said to be "stable." A further approximation of Equation 5.8 can be made if one assumes that the smaller clusters ($1 < j \leq i$) are in equilibrium with the environment for which the second equations have zero as the left side of the equations.

Taking into account the mobility of single atoms and the coalescence of large (immobile) clusters, the above set of kinetic equations will contain the following rates (fluxes):

$$U_j = \sigma_j D_1 n_1 n_j + R a_j n_j,$$

$$U_x = \sigma_x D_1 n_1 n_x + R a_x n_x,$$ (5.9)

$$U_{j^*} = \gamma_{j^*} D_1 n_1^{j^*+1},$$

where:

γ is the evaporation rate of a monomer into the adsorption layer

σ_j are the capture numbers and the first term is proportional to the rate of growth of a j-mer by capturing single atoms due to diffusion in the adsorbed layer

a_j are the areas of clusters with j atoms

R is the rate of the direct impingement of a single atom into the cluster.

By neglecting the second and third terms in Equation 5.8a the equation for the single atoms can be rewritten as follows:

$$\frac{dn_1}{dt} = R - \frac{n_1}{\tau_a} - \left(\sigma_x D_1 n_1 n_x + R a_x n_x \right),$$ (5.10a)

$$\frac{dn_x}{dt} = g_i D_1 n_1^{i+1},$$ (5.10b)

The above parameters σ_x and a_x depend on the cluster shape. If n_x is the average number of atoms in a stable cluster, then the total number of atoms in stable clusters changes with the following rate:

$$\frac{d\left(n_x \bar{n}_x \right)}{dt} = \sigma_x D_1 n_1 n_x + R a_x n_x.$$ (5.11)

For spherical clusters $\bar{n}_x \sim \bar{r}^3$, $a_x \sim \bar{r}^2$, $\sigma_x \sim \bar{r}^p$, where p is a constant. The variables related to the average radius $\bar{r}$ can be obtained

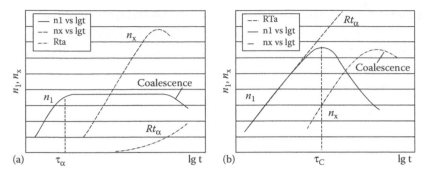

Figure 5.5   Evolution of the monomer density (n_1), stable cluster density (n_x), and the total number of condensed atoms Rt/τ_A, as a function of the deposition time t for two cases: (a) a high temperature, and (b) a low surface temperature.

from three variables: n_1, n_x, and $\bar{r}$. Figure 5.5 shows the characteristic solutions of the Equations 5.10.

Theory by Zinsmeister and Venables was capable of extending the theory to a coating range that was not accessible before. However, this approach is limited to average cluster sizes and therefore was not able to define the distribution function of clusters over sizes.

5.2.3   Numerical methods in the theory of film formation

Adsorption of atoms from gaseous phases was discussed earlier (see Section 1.2). There were also several very detailed reviews available on this topic (see e.g., [63–70]).

The thermodynamic properties of cluster formation were studied by using statistical mechanics methods such as Monte Carlo and MD methods, most of which were collected in the review paper by Abraham [71]. A phase diagram (coexistence curve) of a two-dimensional fluid where atoms were interacting via a Lennard-Jones potential was calculated [70] by using a Monte Carlo method in an isobaric–isothermal ensemble. Figure 1.4 (see Chapter 1) presents such a two-dimensional phase diagram in a coordinate system ρ-T. Such an isobaric-isothermal ensemble was used in [70] to study the melting of a two-dimensional crystalline formed by 575 atoms interacting via a Lennard-Jones potential at a temperature $T = 0.45\ \varepsilon/k_B$ and pressure $P = 0.02\ \varepsilon/\sigma^3$. Here ε is the depth of the potential well and σ is the effective diameter for the interaction between the two atoms. The infinite crystal was simulated by using periodic boundary conditions in two dimensions. The equations of motion were solved by a fifth-order Gear's method with a time increment of 10fs for argon.

The results [72] were presented as a series of snapshots of the basic cell that were scaled to the same cell length. The authors have obtained that the results found by MC and MD are identical. Since MC cannot define the time variable, the MC results were represented as a series of MC steps. Phase diagrams of the two-dimensional Lennard-Jones system were built in the pressure–temperature and temperature–density planes based on the MC calculation results that are presented in Figure 1.6. The obtained results are in good agreement with the liquid-state perturbation theory and for the self-consisted cell theory for the solid state.

The MD method was applied to boiling and evaporating a two-dimensional liquid at $T^*=0.52$ and $P^*=0.01$ where the atoms were interacting via Lennard-Jones potential.* As a result of continuous heating, the MD system size expanded by 20 times after 220,000 time steps and the final state of the system represented a nonideal gas.

In [73], the authors used an MD method for calculating the diffusion coefficient of Xe atoms adsorbed on a Si(100) reconstructed by 2×1 surface at the coating below one monolayer.

The surface temperatures were varied between 40 and 150 K. The surface diffusion coefficient of a single Xe atom adsorbed on the surface was obtained to have a typical Arrhenius form with the activation energy of 2.99 kJ/mol (0.031 eV/atom). One of the advantages of using MD was the non-Markovian character of the atomic jumps in surface diffusion. An analysis of the MC results showed that the jumps of the adatoms are correlated to the interaction between the adatoms and therefore cannot be represented by a Markov chain.

The diffusion coefficient was calculated by the Einstein formula and related the diffusion coefficient with the mean-square displacements of the adatoms:

$$D_S = \lim_{t \to \infty} \frac{\langle x^2(t) \rangle + \langle y^2(t) \rangle}{4t}, \tag{5.12}$$

where:

t is time

---

* The Lennard-Jones units: $T^*=k_B T/\varepsilon$, $P^*=P\sigma^3/\varepsilon$.

⟨⋯⟩ is averaging over all the atoms of the atomic positions on the surface and then over uncorrelated starting points in the simulation:

$$\left\langle x^2(t) \right\rangle = \frac{1}{N_\tau N_p} \sum_{\tau=1}^{N_\tau} \sum^{N_p} [x_p(\tau) - x_p(t+\tau)]^2 , \tag{5.13}$$

where:

N_τ is the number of samplings

N_p of the total number of particles

The results were presented as an Arrhenius formula: $D_s = D_0 \exp(-E_a/k_B T)$ $D_0 = 0.6 \times 10^5$ cm^2/s, and $E_a = 2.99$ kJ/mol for the temperature interval between 40 and 150 K, and surface coverage value of $\theta = 0.6$.

References

1. R. Lodis and R. Parker, *Growth of Single Crystals*, Mir: Moscow, 1974.

2. R. A. Haefer, *Cryopumping: Theory and Practice*, p. 157, Clarendon Press: England, 1989.

3. T. Takagi, *Ionized Cluster Beam Deposition and Epitaxy*, vol. 1, Noyes Publications: Park Ridge, NJ, 1988.

4. Yu. K. Tovbin, *Theory of Physical Chemistry Processes at a Gas-Solid Interface*, CRC: Boca Raton, FL, 1991.

5. U. Leuthäusser, Kinetic theory of adsorption and desorption, *Z Phys B* 44, 101–108, 1981.

6. F. O. Goodman and H. Y. Wachman, *Dynamics of Gas: Surface Scattering*, p. 327, Academic: New York, 1976.

7. R. Oman, A. Bogan, C. Weiser, and C. Li, Interactions of gas molecules with an ideal crystal surface, *AIAA J* 2, 1722–1730, 1964.

8. R. A. Oman, Numerical experiments on scattering of noble gases from single-crystal silver, *J Chem Phys* 48, 3919, 1968.

9. Fifth International Symposium on Small Particles and Inorganic Clusters (ISSPIC 5), Conference Abstracts, University of Konstanz, Germany, 1990.

10. P. Jena. B. K. Rao, S. N. Khanna, and D. Reidel (eds), *Physics and Chemistry of Small Clusters*, NATO ASI Series B Physics, vol. 158, Plenum Press: New York, 1987.

11. P. Jena, S. N. Khanna, and B. K. Rao (eds), *Physics and Chemistry of Finite Systems: From Clusters to Crystals*, NATO ASI Series, vols. 1 and 2, Kluwer Dordrecht, 1992; Sec. C: *Math Phys Sci* 374.

12. P. Jena and S. N. Behera, Clusters and nanostructured materials, in *Proceedings of the International Workshop on Clusters and Nanostructured Materials* (Puri, India), Nova Science: New York, 1996.

13. J. Jellinek (ed.), *Theory of Atomic and Molecular Clusters*, in R. S. Berry, A. W. J. Castleman, H. Haberland, J. Jortner, and T. Kondow (eds), Springer Series in Cluster Physics, 1999.

14. Y. Kawazoe, T. Kondow, and K. Ohno (eds), *Clusters and Nanomaterials: Theory and Experiment*, in R. S. Berry, A. W. J. Castleman, H. Haberland, J. Jortner, and T. Kondow (eds) Springer Series in Cluster Physics, 2002. Series: Springer Series in Cluster Physics.

15. R. M. Logan and J. C. Keck, Classical theory for the interaction of gas atoms with solid surfaces, *J Chem Phys* 49(2), 860, 1968.

16. M. M. Eisenstadt, Condensation of gases during cryopumping: The effect of surface temperature on the critical energy of trapping, *J Vac Sci Techn* 78(4), 479–484, 1970.

17. A. Hurkmans, E. G. Overbosch, D. R. Olander, and J. Los, The trapping of potassium atoms by a polycrystalline tungsten surface, *Surf Sci* 54(1), 154–168, 1976.

18. U. Leuthäusser, Thermal desorption of atoms from surface, *Phys Rev B* 36(9), 4672–4680, 1987.

19. A. P. Kryukov, One-dimensional steady condensation of vapor velocities comparable to the velocity of sound, *Fluid Dyn* 20(3), 487–491, 1985.

20. V. I. Zhuk, Capture of gas atoms on a solid surface, *J Appl Mech Techn Phys* 20(1), 4–7, 1979.

21. S. A. Adelman and J. D. Doll, Generalized Langevin equation for atom/solid-surface scattering: Collinear atom/harmonic chain model, *J Chem Phys* 61, 4242–4245, 1974.

22. J. D. Doll, L. E. Myers, and S. A. Adelman, Generalized Langevin equation approach for atom/solid–surface scattering: Inelastic studies, *J Chem Phys* 63, 4908, 1975.

23. J. D. Doll and D. R. Dion, Generalized Langevin equation approach for atom/solid–surface scattering: Numerical techniques for Gaussian generalized Langevin dynamics, *J Chem Phys* 65, 3762, 1976.

24. J. C. Tully, Dynamics of gas–surface interactions: 3D generalized Langevin model applied to fcc and bcc surfaces, *J Chem Phys* 73, 1975, 1980.

25. M. Shugard, J. C. Tully, and A. Nitzan, Dynamics of gas–solid interactions: Calculations of energy transfer and sticking, *J Chem Phys* 66, 2534, 1977.

26. G.-Q. Xu, R. J. Holland, S. L. Bernasek, and J. C. Tully, Dynamics of cluster scattering from surfaces, *J Chem Phys* 90, 3831, 1989.

27. C. W. Muhlhausen, L. R. Williams, and J. C. Tully, Dynamics of gas–surface interactions: Scattering and desorption of NO from Ag(111) and Pt(111), *J Chem Phys* 83, 2594–2606, 1985.

28. R. R. Lucchese and J. C. Tully, Trajectory studies of vibrational energy transfer in gas–surface collisions, *J Chem Phys* 80, 3451, 1984.

29. C. R. Arumainayagam, R. J. Madix, M. C. McMaster, V. M. Suzawa, and J. C. Tully, Trapping dynamics of xenon on Pt(111), *Surf Sci* 226, 180–185, 1990.

30. E. K. Grimmelmann, J. C. Tully, and E. Helfand, Molecular dynamics of infrequent events: Thermal desorption of xenon from a platinum surface, *J Chem Phys* 74, 5300, 1981.

31. R. J. Holland, G. Q. Xu, J. Levkoff, A. Robertson, and S. L. Bernasek, J. Experimental studies of the dynamics of nitrogen van der Waals cluster scattering from metal surfaces, *Chem Phys* 88, 7952, 1988.

32. T. Takagi, I. Yamada, M. Kunori, and S. Kobiyama, Vaporized metal cluster ion source for ion plating, in *Proceedings of the 2nd International Conference on Ion Sources*, p. 790, Österreichishe Studiengessellschaft für Atomenergie, Vienna, 1972.

33. T. Takagi, I. Yamada, K. Yanagawa, M. Kuniri, and S. Kobiyama, Vapourized-metal cluster ion source for ion plating, *Japan J Appl Phys Suppl* 2(1), 427–430, 1974.

34. T. Takagi, I. Yamada, and A. Sasaki, An evaluation of metal and semiconductor films formed by ionized-cluster beam deposition, *Thin Solid Films* 39, 207–217, 1976.

35. T. Takagi, I. Yamada, and A. Sasaki, Ionized-cluster beam deposition and epitaxy as fabrication technique for electron devices, *Thin Solid Films* 45, 569–576, 1977.

36. T. Takagi, Role of ions in ion-based film formation, *Thin Solid Films* 92, 1–17, 1982.

37. T. Takagi and I. Yamada, Ionized-cluster-beam deposition of optical interference coatings, *Appl Optics* 24, 879–882, 1985.

38. T. Takagi, Ionized cluster beam (ICB) deposition and processes, *Pure Appl Chem* 60(5), 781–794, 1988.

39. H. Haberland, Z. Insepov, M. Karrais, M. Mall, M. Moseler and Y. Thurner, Thin-film growth by energetic cluster impact (ECI)—Comparison between experiment and molecular-dynamics simulations, *Mater Sci Eng B* 19, 31–36, 1993.

40. H. Haberland, Z. Insepov, and M. Moseler, Molecular-dynamics simulation of thin-film formation by energetic cluster impact (ECI), *Z Phys D* 26, 229–231, 1993.

41. H. Haberland, Z. Insepov, M. Karrais, M. Mall, M. Moseler and Y. Thurner, Thin-films from energetic cluster-impact: Experiment and molecular-dynamics simulations, *Nucl Instrum Methods Phys Res B* 80(1), 1320–1323, 1993.

42. H. Haberland, Z. Insepov, and M. Moseler, Molecular-dynamics simulation of thin-film growth by energetic cluster-impact, *Phys Rev B* 51, 11061–11067, 1995.

43. K. -H. Müller, Cluster-beam deposition of thin films: A molecular dynamics simulation, *J Appl Phys* 61, 2516–2521, 1987.

44. R. Biswas, G. S. Grest, and C. M. Soukolis, Molecular dynamics simulations of cluster and atom deposition on silicon (111), *Phys Rev B* 38, 8154–8162, 1988.

45. I. Kwon, R. Biswas, G. S. Grest, and C. M. Soukoulis, Molecular dynamics simulation of amorphous and epitaxial Si film growth on Si(111), *Phys Rev B* 41, 3678–3687, 1990.

46. H. Hsie and R. S. Averback, Molecular dynamics simulations of collisions between energetic clusters of atoms and metal substrates, *Phys Rev B* 45, 4417–4430, 1992.

47. M. H. Shapiro and T. A. Tombrello, Simulation of core excitation during cluster impacts, *Phys Rev Lett* 68, 1613–1615, 1992.

48. Y. Yamamura, I. Yamada, and T. Takagi, Computer studies of ionized cluster beam deposition, *Nucl Instrum Methods B* 37/38, 902–905, 1989.

49. Y. Yamamura, Computer simulation of ionized cluster beam bombardment on a carbon substrate, *Nucl Instrum Methods B* 45, 707–713, 1990.

50. Y. Yamamura, Energy distribution of constituent atoms of cluster impacts on solid surface, *Nucl Instrum Methods B* 62, 181–190, 1991.

51. P. Sigmund and C. Claussen, Sputtering from elastic-collision spikes in heavy-ion-bombarded metals, *J Appl Phys* 52, 990–993, 1981.

52. H. Urbassek and P. Sigmund, A note of evaporation from heated spikes, *Appl Phys A* 35, 19–25, 1984.

53. D. A. Thompson, Application of an extended linear cascade model to the sputtering of Ag, Au and Pt by heavy atomic and molecular ions, *J Appl Phys* 52, 982–989, 1981.

54. R. Beuhler and L. Friedman, Larger cluster ion impact phenomena, *Chem Rev* 86, 521–537, 1986.

55. I. Yamada, W. L. Brown, J. A. Northby, and M. Sosnowski, Surface modification with gas-cluster-ion beams, in *Proceedings of the International Conference on Appl Accel in Res and Industry*, University of North Texas, USA, 1992.

56. A. Wucher, M. Wahl, and H. Oechsner, Sputtered neutral silver clusters up to Ag18, *Nucl Instrum Methods Phys Res B* 82, 337–346, 1993.

57. W. Husinsky, G. Nicolussi, and G. Betz, Energy distribution of sputtered metal Al-clusters, *Nucl Instrum Methods Phys Res B* 82, 323–328, 1993.

58. S. R. Coon, W. F. Calaway, M. J. Pellin, G.A. Curlee, and J. M. White, Kinetic energy distributions of sputtered of sputtered neutral aluminium clusters: Al-Al6, *Nucl Instrum Methods Phys Res B* 82, 329–336, 1993.

59. J.-O. Bovin and J.-O. Malm, Atomic resolution electron microscopy of small metal clusters, *Z Phys D* 19, 293–298, 1991.

60. E. E. Polymeropoulos and J. Brickmann, Molecular dynamics study of the formation of argon clusters in the compressed gas, *Chem Phys Lett* 92, 59–63, 1982.

61. E. E. Polymeropoulos, P. Bopp, J. Brickmann, L. Jansen, and R. Block, Molecular-dynamics simulations in systems of rare-gases using Axilrod-Teller and exchange three-atom interactions, *Phys Rev A* 31, 3565–3569, 1985.

62. A. A. Belov and Yu. E. Lozovik, Molecular dynamic study of cluster formation in gases, *Preprint of Inst Spectroscopy of RAS* 159, 1–40, 1988.

63. J. A. Venables, Rate equation approaches to thin film nucleation kinetics, *Philos Mag* 27(3), 697–738, 1973.

64. F. F. Abraham, On the thermodynamics, structure and phase stability of the nonuniform state, *Phys Rep* 59, 92–156, 1979.

65. T. N. Rhodin and D. Walton, *Metal Surfaces*, p. 259, American Society for Metals: Cleveland, 1963.

66. J. A. Venables, G. D. T. Spiller, and M. Hanbücken, Nucleation and growth of thin films, *Rep Prog Phys* 47, 399–459, 1984.

67. G. Zinsmeister, A contribution to Frenkel's theory of condensation (Part A), *Vacuum* 16, 529–535, 1966.

68. G. Zinsmeister, Theory of thin film condensation: Solution of the simplified condensation equation (Part B), *Thin Solid Films* 2, 497–507, 1968.

69. G. Zinsmeister, Theory of thin film condensation: Aggregate size distribution in island films (Part C), *Thin Solid Films* 4, 363–386, 1969.

70. G. Zinsmeister, Theory of thin film condensation: Influence of a variable collision factor (Part D), *Thin Solid Films* 7, 51–75, 1971.

71. A. A. Chernov, *Modern Crystallography III*, Springer Series in Solid-State Sciences, vol. 36, Springer: Berlin, 1984.

72. F. F. Abraham, Phases of two-dimensional matter, *Phys Rep* 80, 339–374, 1981.

73. L. Utrera and R. Ramirez, Molecular dynamics simulation of Xe diffusion on the Si(100)-2x1 surface, *J Chem Phys* 96, 7838–7847, 1992.

CHAPTER 6

Cluster Formation Kinetics in Thin Film Growth*

6.1 Introduction

Sticking gas particles, such as atoms, molecules, or multiatomic clusters, result in the formation of a thin film on the surface of a solid polycrystalline substrate. The molecular dynamics (MD) methods discussed in previous chapters allow one to study the kinetics of cluster formation, an initial stage of thin film growth for the Volmer–Weber model [1–6].

In this chapter, MD is used to calculate the kinetic coefficients for the growth of surface islands of a new phase (Section 6.2). The obtained coefficients are used in the next few sections for solving the microscopic kinetic equations of cluster growth (Section 6.3) for rarefied absorption layers. A two-dimensional (2D) surface density of adsorbed atoms is introduced as follows: $n = N/N_0$, where N is the number of adatoms and N_0 is the number of possible absorption sites. Another name for 2D density is coverage.

Since the kinetic rate equations based on mean-field theory are not applicable for a dense absorption layer, MD is the only method that enables one to study cluster formation and obtaining the densities of small clusters. This will be discussed in Section 6.4.

If a dense gas residing in a thermodynamically equilibrium and stable state is abruptly cooled down (quenched) so that temperature in the system becomes much lower than the critical one for the gas–liquid phase transformation, the state of the gas system becomes unstable and the system spontaneously transforms into a two-phase system that can be characterized by a typical size that is determined by the system's physical properties. This process is called spinodal decomposition. We discuss comparison with experiment in Section 6.5, spinodal decomposition in Section 6.6, modeling of cluster evaporation in Section 6.7, and phase transition in clusters in Section 6.8.

---

* The contents of this chapter have been published in [13,16,18,22].

6.2 Calculation of Diffusion Coefficients

A typical premise about the immobility of the small islands, the clusters of the new phase in the adsorption layers, was proved to be incorrect in a series of papers by Kinoshita and coworkers and Harsdorff and Reiners [7–10]. Since the small islands are the most mobile species, they can significantly change the kinetics of thin film nucleation and growth for cases where the mechanism of Volmer–Weber (VW) is applicable [11]. Experimental observation of small clusters is difficult since electron microscopy cannot detect species that are smaller than the wave length of the electrons used for detection [12].

The diffusion coefficients of small Ar and Pt clusters, with sizes from 1 to 6 atoms, deposited on the top of a graphite basal surface were calculated by MD in [13]. The simulation model was described in detail in Chapter 2 and it therefore will only be discussed briefly here.

The simulation models that are based on the stochastic dynamics of adatoms usually use the friction coefficient γ via a Debye surface frequency ω_D of the substrate [14]. However, for the simulation model used here, a better choice of the friction constant is the expression given in [15]:

$$\gamma = \tau_E^{-1} = \frac{4\pi\lambda d}{c_V},$$ (6.1)

where:

τ_E is the relaxation time of an excess energy of a small volume, with a characteristic size d

λ is the thermal conductivity

c_V is the specific energy in the real system

The numerical factor in Equation 6.1 corresponds to the hemispherical shape of the cluster. The parameter γ determines the frequency spectrum of the velocity fluctuations in the system [16]. In the case of a single adatom, the parameter g is determined from the Einstein expression:

$$\gamma = \frac{k_B T}{m D_1},$$ (6.2)

where D_1 is the diffusion coefficient of a single adatom on the surface, which is equal to the low-frequency limit of the velocity auto-correlation function [2]. More detail on this model can also be found in Chapter 2 where the molecular models are discussed.

Expression Equation 6.1 was obtained for an infinite media. Since our model system contains only a small number of atoms, n, the friction coefficient can be determined as an interpolation between that of a single adatom and that obtained from Equation 6.1, which is equal to $\gamma(1) = \pi\omega_D/6$, at $n=1$ and to Equation 6.1 when the number of atoms becomes very large:

$$\gamma(n) = \gamma(\infty) + \left[\gamma(1) - \gamma(\infty)\right] \cdot n^{-2/3}. \tag{6.3}$$

A canonical ensemble of clusters of a certain size is formed and the time-dependent dynamics of the clusters is studied by MD. At large time scales, $t \gg \gamma^{-1}$, all cluster energy components are defined by the stationary (independent of time) values:

$$\left\langle E_t \right\rangle = \frac{3}{2} k_B T$$

$$\left\langle E_r \right\rangle = k_B T \tag{6.4}$$

$$\left\langle E_v \right\rangle = (3n - 5) k_B T,$$

where:

E_t, E_r, and E_v correspond to the translational, rotational, and vibrational components of the total cluster energy

$\langle ... \rangle$ is the averaging of the energy components over time at the stationary stage of the system

The accuracy of the velocity distribution of the atoms in the cluster generated by such a procedure can be verified by controlling their proximity to a Maxwellian function that was proposed by Rahman [17]. During the calculation, the average square root of the center of mass of the cluster location $\left\langle \Delta r_{c.m.}^2 \right\rangle$ was calculated and the diffusion coefficient D_n was obtained on the linear stage using the formula

$$D_n = \lim\left(\frac{\left\langle \Delta r^2 \right\rangle}{4t}\right), \quad t \gg \gamma^{-1}.$$

Figure 6.1 illustrates the dependence of translational, rotational, and vibrational energies of a Pt dimer, Pt_2, on the time elapsed from the beginning of motion, for $\gamma = 1$ and 3.

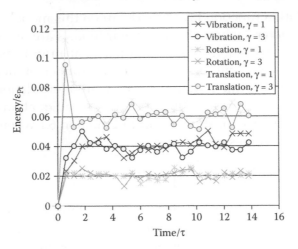

Figure 6.1   Dependencies of vibrations, rotations, and translations Pt_2 dimer energies on a basal graphite surface on time; $\gamma=1$ (x), $\gamma=3$ (●).

It is shown that the dimer Pt_2 quickly (within 100 time steps) obtains thermodynamic equilibrium with the thermostat (substrate) and diffuses in the stationary stage of motion. The distribution function becomes Maxwellian within the same period of time.

Figure 6.2 shows the dependence of the diffusion constant on the cluster size ($n=1, 2, 4, 6$ are the numbers of atoms in clusters) at room temperature $T_s=300$ K.

Figure 6.3 depicts the dependence of the diffusion constant of platinum clusters Pt_n in the cluster size. Line 1 corresponds to the values calculated in our paper [18] where the diffusion coefficient of a single adatom is equal to $D_1=2\times10^{-5}$ cm^2/s, which is a reasonable value. Experimental data on the Pt adatom diffusion on the surface are missing. However, the diffusion coefficient of a similar system where Pt atoms were deposited on a NaCl surface was obtained to be $D_1=5\times10^{-6}$ cm^2/s, which is close to our MD prediction. The same authors noted that the diffusion coefficient D_1 of the Pt/graphite system should be larger than that of Pt/NaCl.

Figure 6.3 shows that the dependence of the diffusion constant D_n for the Pt/graphite system on the cluster size n can be fitted well with the dependence proposed by Kashchiev [19]:

$$D_n = D_1 \cdot n^{-\alpha}, \tag{6.5}$$

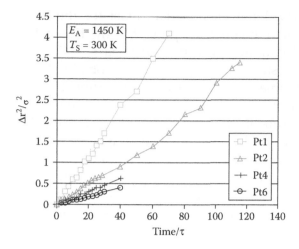

Figure 6.2 Mean-square displacements of platinum clusters, consisting of 1, 2, 4, and 6 atoms, on a basal graphite surface obtained by MD at $T_S = 300$ K.

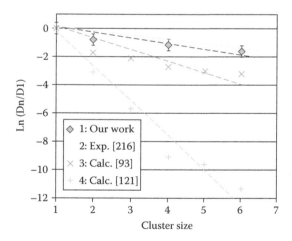

Figure 6.3 Dependencies of the diffusion coefficient D_n of platinum clusters of size n. 1—our calculations; 2—experimental data for Au/NaCl [20]; 3, 4—calculations [12,21]. (J. C. Tully, *J Chem Phys* 73, 1975–1986, 1980; J. A. Venables, et al., *Rep Prog Phys* 47, 399–459, 1984; A. Y. Valuev, et al., *J Phys Chem* 63, 1469–1475, 1989.)

where $\alpha = 0.87$. Experimental data exist for the Au/NaCl system only [12], where the power exponent in the Kaschchiev expression Equation 6.5 parameter was measured as $\alpha = 0.67$.

Figure 6.3 shows the dependence D_n obtained in [20] by adapting the experimental data where $\alpha = 0.67$ (line 2). Lines 3 and 4 are the experimental diffusion data for D_n obtained in [12,21]. It shows that the experimental dependence 2 is close to our calculated results shown as line 1. The reason for such agreement comers from the same nature of the diffusion in our simulation and experiment, namely, we simulate week adhesion, where the clusters have a hemispherical shape. As such the number of atoms touching the substrate n_s is proportional to the total number in a 2/3 power: $n_s \sim n^{2/3}$.

We will show here that in this case the diffusion constant should have the typical dependence $D_n \sim n^{-2/3}$. If the friction constant of one adatom is γ, the diffusion coefficient of a single adatom can be expressed as $D_1 \approx k_B T / m\gamma$.

The cluster velocity can be determined as $v_i \approx v_{c.m.} + \delta v_i$, where $v_{c.m.}$ is the center-of-mass velocity of the cluster, δv_i is the velocity of the adatom relative to the center of mass of the cluster. In this case, we can get the following expression:

$$m\dot{v}_i = m\dot{v}_{c.m.} + m\delta\dot{v}_i = -\gamma m v_{c.m.} - \gamma m \delta v_i + R_i.$$

By summation over all n_s adatoms touching the substrate, we get that at $n_s \gg 1$:

$$\Gamma = \frac{\gamma n_s}{n}, \tag{6.6}$$

where Γ is the friction constant for the whole cluster. By replacing Γ in the Einstein equation by expression Equation 6.6, we get:

$$D_n = \frac{k_B T}{M\Gamma} = \frac{k_B T}{\gamma m n_s} = D_1 \cdot n^{-2/3}.$$

Therefore, it seems that the value $\alpha = 2/3$ is a characteristic of the hemispherical shape of the cluster. The power exponent $\alpha > 1$ characterizes a flat geometry of the cluster on the surface and an "intermediately strong" adhesion.

6.3   Solutions of Kinetic Equations

The following kinetic equations for cluster densities (concentrations) were solved:

$$\frac{dn_1}{dt} = -\sum_{i=1}^{\infty} c_{1i} n_1 n_i,$$

$$\frac{dn_i}{dt} = \sum_{j+k=i}^{\infty} c_{jk} n_k n_j - \sum_{j=1}^{\infty} c_{ij} n_i n_j.$$

(6.7)

The kinetic coefficients c_{ij} were obtained using the empiric relation $c_{ij} = (\sigma_i + \sigma_j) D_{ij}$, where σ_{ij} are the cluster diameters of the clusters. D_{ij} is the diffusion coefficient.

In Equation 6.7 the evaporation of the cluster was neglected which is acceptable for surfaces at room temperature. The diffusion coefficient was calculated previously in Section 6.1.

The numerical solution of the set of Equation 6.7 was carried out on a digital computer with 30 kinetic equations in the set. The surface density of the adsorbed atoms corresponded to the condensation of Pt on graphite and Pd on MgO ($\gamma^* = 0.05$).

In Figure 6.4 there are some characteristic solutions of the kinetic Equations 6.7 for four time instants (in reduced units).

6.4 Molecular Dynamics of Cluster Formation on Surfaces

The simplest representation of the adsorption layer is a 2D model consisting of N particles interacting via a certain pair potential that

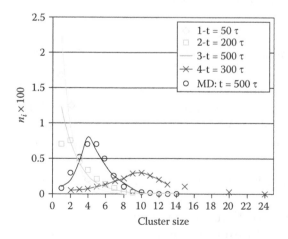

Figure 6.4 Evolution of the cluster size distribution function in time. Comparison of the solution of kinetic Equation 6.7 with MD calculations at the same time instant 500 τ.

are moving within an {x,y} plane. Clusters in such two-dimensional models will be disks and the interaction of clusters with the substrate in such a simplified system is neglected. A more realistic three-dimensional model includes interaction of the adsorbate atoms with the substrate atoms, and clusters in such three-dimensional models will have three-dimensional shapes.

Although it is less realistic for most strongly interacting adsorbate atoms with surfaces, the two-dimensional model is adequate for systems where the adsorbate atoms are weakly interacting with the smooth substrates, such as atoms of noble elements on graphite or pyrolytic carbon surfaces.

In thin film formation technology, such processes are called a *layer-by-layer* condensation mechanism, where the islands of the new deposited phase have a two-dimensional structure, and they by attaching the new adatoms from the same adsorbate layer, until they fill out the adsorbate layer, and the next layer starts only after the atoms complete the layer below.

In a mixed case called the Frank–van der Merwe mechanism, the growth of the film starts by building two-dimensional islands on the substrate, and the formation of three-dimensional islands starts on the completed two-dimensional atomic layers.

Therefore, although it is simplified, the two-dimensional model is useful for a number of adsorbate systems where the adsorption energy is much less than the interaction between the adsorbate atoms, and also to develop useful qualitative understandings [13,16,18,22].

Structural changes in such systems can be characterized by the time-dependent structural factor, $S(k\lambda, t)$, where $k\lambda$ is the wave number.

Figure 6.5 shows the evolution of a cluster distribution in a two-dimensional system, after 4000 time steps, for several simplest models: dash-dot curve 1—a two-dimensional model; curve 2—an isothermal MD model, where adsorbate atoms interact with the substrate; curve 3—a Brownian dynamics model. In this figure, the histogram shows the experimental data obtained in [23].

The initial positions and velocities were selected as follows: 504 atoms were uniformly distributed at an average density $\rho\sigma^2 = 0.13$ and temperature $k_BT/\varepsilon = 0.04$. In the following, time will be measured in Lennard-Jones units, $\tau = \sigma(m/\varepsilon)^{1/2}$. A time increment was chosen $\Delta t = 0.04\ \tau$.

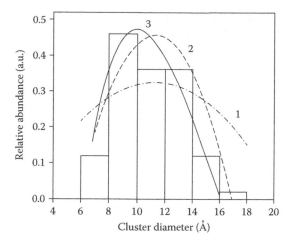

Figure 6.5 Distribution functions of Pt clusters on a basic graphite plane: 1, calculated by a two-dimensional MD after 4000 time steps; 2, isothermal MD with interaction with the substrate; 3, Brownian dynamics; histogram, experiment. (From J. F. Hamilton, et al., *Surf Sci* 106, 146–151, 1981.)

For comparison with experiment, simulation should take into account a realistic description of the surface. Calculations for a three-dimensional model taking into account surface structure and interaction of adsorbed atoms with the substrate were carried out for the deposition of Pt vapor into a basal graphite surface in conditions identical to the experimental setup used in [23] and Pd vapor on a MgO (100) surface described in [24].

A graphite surface was built as a perfect hexagonal foil structure, with the nearest neighbor's distance in the lattice at 1.42 Å [25]. Interactions between two Pt atoms were modeled by the Lennard-Jones potential (see, e.g., Equation 2.35), with the following parameters [26]: $\varepsilon_{Pt} = 7910$ K, $\sigma_{Pt} = 2.54$ Å.

The parameters for interactions between Pt and carbon atoms were modeled by the Lorentz–Berthelot rule: $\varepsilon_{Pt-C} = \sqrt{\varepsilon_{Pt}\varepsilon_C}$ and $\sigma_{Pt-C} = (\sigma_{Pt} + \sigma_C)/2$, where $\varepsilon_C = 29$ K and $\sigma_C = 3.4$ Å [25]. The interaction potential was cut off at a distance of $R_c = 2\sigma_{Pt}$.

A similar model was used for simulating adsorbed atoms of Pd in a MgO surface, and the charge-neutral atom interaction potential was chosen as

$$U(r) = 4\varepsilon' \left[\left(\frac{\sigma'}{r} \right)^{12} - \left(\frac{\sigma'}{r} \right)^{4} \right]. \tag{6.8}$$

The MgO crystalline structure was a simple cubic with the following ionic radii: 0.65 Å for Mg^{2+} and 1.46 Å for O^{2-} [27]. The specific values of σ' were obtained from the ionic radii of the Pd atomic diameter, 2.6 Å, by using the Lorentz–Berthelot rule. $\varepsilon_{Pd} = 5000$ K was estimated by assuming that the ratio of the potential depth wells, ε_{Pd} for Pd and ε_{Pt} for Pt, is approximately equal to the ratio of sublimation energies for these two metals, which are given in [27]. Interactions with the substrate atoms beyond R_c were taken into account in a continuum approximation for a model Lennard-Jones potential. ε' in expression Equation 6.8 was obtained by a separate MD code by calculating the desorption energies of single Pd atoms on a MgO surface at an experimental temperature and further tuning it to the experimental value of 5000 K [28]. Averaging over 100 samples gave the following value: $\varepsilon' = 250$ K.

Calculation of the friction coefficient was carried out by Equation 2.35. The diffusion activation energy was chosen for Pt atoms on a graphite basal surface: $E_a/\varepsilon_{Pt} = 0.2$ at $T^* = 0.04$, and for Pd on a MgO surface as $E_a/\varepsilon_{Pd} = 0.6$ at $T^* = 0.08$.

Frequencies of atomic vibrations were obtained according to

$$\omega_s = v_0 \left(\frac{6\pi^2}{\Omega_0} \right)^{1/3},$$

where:

$v_0 = [1/3(1/v_l^3 + 2/v_t)] - 1/3$ (v_l and v_t are the longitudinal and transversal components of the sound speed)

Ω_0 is the volume of the basic cell

Estimating the Pt friction constant gives $g = 0.1$—1 (in τ^{-1} units). An estimate of the friction constant of Pd on a MgO surface was carried out from $\omega_s = 13 \times 10^{13}$ 1/s, which gives $\gamma = 0.1$.

6.5   Comparison with Experiment

Figure 6.5 shows comparisons between MD simulation results for three different model systems of thin film nucleation on an atomically flat solid surface: a two-dimensional model, a three-dimensional isothermal model that includes the interaction of adsorbed layer atoms with ones belonging to the surface, and a three-dimensional Brownian dynamics model. The calculation parameters correspond to the experimental data obtained from [23]: $k_B T/\varepsilon_{Pt} = 0.04$, $\rho \sigma_{Pt}^2 = 0.13$.

The cluster diameters were estimated according to the method proposed in experiment [23] where the cluster size was obtained by the area covered by the cluster. Curve 2 was obtained after 10,000 time steps, with increment of $D_t = 0.03\tau$, and $N = 200$, and curve 3 was obtained after 1400 time steps, with the model in which a Langevin force was acting on the atoms of the first layer that were in close proximity of the surface ($z_i < \sigma_{Pt}$).

A similar calculation result shown in Figure 6.6 compares our calculations with experiment on the deposition of Pd atoms on a (100) surface of MgO at a surface temperature of $T = 400$ K [28].

Curve 1 was calculated by a two-dimensional MD model where the surface atoms were not interacting with the cluster atoms. Curve 2 was calculated according to three-dimensional isothermal Brownian dynamics where cluster atoms were interacting with the substrate via Langevin forces, after 24,000 time steps, with increment of $\Delta t = 0.02$, $\gamma = 0.05\,\tau_{-1}$, $\rho^* = 0.05$, $T^* = 0.08$, $N = 200$.

Cluster formations occur with different rates and reach different stationary levels. In an isothermal ensemble, clusters have more choices to coagulate (assemble) into larger clusters, and therefore they have higher temperatures. Additionally, this ensemble allowed them to attain a hydrodynamic velocity.

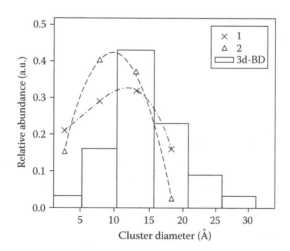

Figure 6.6 Comparison of calculated Pd cluster size distribution functions on a (100) MgO surface with experiment in 1, 2 calculated by MD: 1, two-dimensional model; 2, three-dimensional isothermal MD with Brownian dynamics; histogram, experiment. (From C. Chapon, et al., *Surf Sci* 162, 747–754, 1985.)

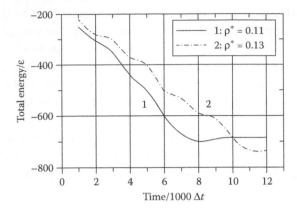

Figure 6.7　Total energy of adsorption layer in three-dimensional isothermal ensemble: 1—$\rho^* = 0.11$; 2—$\rho^* = 0.13$.

Reaching a stationary state over cluster sizes was also controlled by a dependence of the total cluster energy on time. Figure 6.7 shows that after 12,000 time steps in an isothermal ensemble, the system attains a stationary state. A similar behavior shows the three-dimensional Brownian dynamics model.

The following remarks can be made: cluster size distribution does not depend on the value of γ_0; this parameter determines the speed with which the clusters reach their stationary size distributions. A similar behavior was observed in the time dependence of the total energy of the system. A cluster size distribution function in a two-dimensional system shows a nonstationary character, since it constantly moves to larger cluster sizes.

Figures 6.8 and 6.9 show the structure of islands in the direction perpendicular to the surface for Pd/MgO and Pt/graphite systems. They show that the island consists of at least two or three layers. It is known [1] that such a shape of islands (clusters) is realized in case of weak adhesion of the adsorbed atoms to the substrate.

6.6　Spinodal Decomposition of a Supersaturated Adsorbed Atomic Layer

In a specific case of weakly adsorbed layers, the interaction between the atoms inside the layer is much stronger than that between the atoms and the substrate: $\varepsilon_{Pt\text{-}Pt} \gg \varepsilon_{Pt\text{-}C}$. In this case, the dynamics of the adsorbed layer can be very efficiently determined by using the

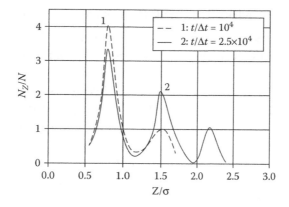

Figure 6.8   Dependence of Pd adatoms on the distance to surface for models with Brownian dynamics ($\gamma_0 = 0.05\ \tau^{-1}$): $t/\Delta t = 10^4$, $1-t/\Delta t = 2.5 \times 10^4$.

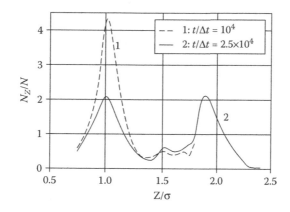

Figure 6.9   Dependence of the Pt adatom density on the distance to surface for an isothermal ensemble: (1) $t/\Delta t = 10^4$, (2) $t/\Delta t = 2.5 \times 10^4$.

concept of spinodal decomposition. The characteristic wave length of the density wave (1.15) for the fastest growth will be a typical size of the clusters in the system. In this case, for the wave length λ_m for the system undergoing spinodal decomposition at $T^* = 0.04$ and having density $\rho^* = 0.13$, the wave length can be calculated by the following approximation: $\lambda_m = 2\pi/k_m \approx 4$. The average area occupied by one particle is equal to $7\sigma^2$. Therefore, the average cluster size is estimated to be $\bar{k} \approx 2.3$. This estimate agrees with the MD result giving several atoms in the cluster.

6.7   Modeling of Cluster Evaporation

The molecular dynamics method provides a simple way to investigate another important thermal process—evaporation of a cluster, or cluster decay, where the cluster transforms into a system of separate atoms on the surface. An MD model consists of a small cluster placed on the surface and heated up to temperature $T^*=0.4$. The friction constant in the equations of motion will changes according to formula (6.2).

Figure 6.10 shows the results of such a simulation for the evolution of the cluster size during heating of the cluster on a graphite surface. An analysis of the trajectories shows that two processes are going on during the cluster heat-up and decay: (1) separation of large clusters into small ones and evaporation of small clusters, and (2) coagulation of small clusters into large ones. The last process is dominated by the higher mobility of small clusters. The size distribution function is smeared up, showing the formation of both large and small clusters.

Figure 6.11 shows the dependence of the particle density on the distance along the normal to the surface, after the system reaches an equilibrium state. As this figure shows, a laminar structure of the clusters is replaced by a rough adatomic structure. In addition, one can see a new evaporation event where a cluster consisting of 6–7 atoms was evaporated from the surface as a whole.

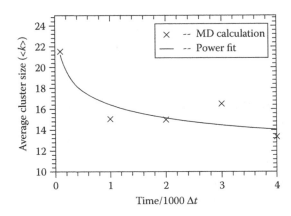

Figure 6.10   Evolution of the average cluster size during substrate heating ($\rho^*=0.13$, $T_s^*=0.4$, $\gamma_0=0.3$).

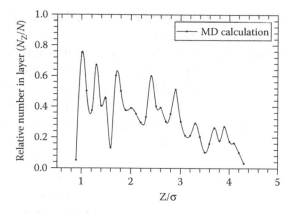

Figure 6.11 Distribution of adatoms along the normal to the surface for the parameters $\rho^*=0.13$, $T_s^*=0.4$, $\gamma_0=0.3$.

6.8 Modeling of Phase Transition Leading to Formation of a Wetting Layer of the Surface

Figure 6.12 shows an initial structure of an adsorbed layer that represents a dense packed system of 90 adatoms (10 layers with 9 atoms in each) and a cluster on a graphite surface.

As an adsorbate, molecules of C_2H_4 and CF_4 were selected for which there were available experimental data for film growth according to the Stranski–Krastanov mechanism. Modeling of adatom evaporation into adsorbed films was carried out for two cluster sizes: $k=83$ (37, 27, and 19 atoms in 1–3 layers) and $k=38$ (19, 12, and 7 atoms in 1–3 layers).

The adatoms were interacting via a Lennard-Jones potential with the following parameters: ethylene (C_2H_4) ($\varepsilon=224$ K, $\sigma=4.16$ Å) and tetrafluoride (CF_4) ($\varepsilon=134$ K, $\sigma=4.7$ Å) [29]. The interaction parameters between adatoms and surface atoms were obtained according the Lorentz–Berthelot mixing rule (cf. Section 2.9). The

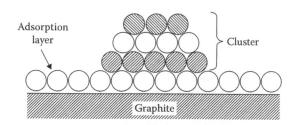

Figure 6.12 Initial structure of cluster and adsorption layer.

boundary conditions were selected as free on the top and reflecting at the bottom of the simulating system.

The modeling system was equilibrated at low temperatures, so that the structure of the cluster did not change during the equilibration, during several hundred time steps, and then suddenly was raised up to the reference temperature (from 0.35 to 0.6 in reduced units). The evolution of the system was observed within 2000 time steps, where the time step was 0.01.

The rate of cluster decay was controlled by calculating the probability ν of transition of a cluster atoms into an adsorbed layer on the surface, that is, separation of atoms from the cluster and to be able to freely move on the surface far from the cluster.

In [30–32] (see also [33–34]), this process was obtained experimentally to be a phase transition of the first kind.

6.8.1   Comparison with experiment

Figure 6.13 shows the dependence of the probability of transition (separation) of atoms from the cluster into an adsorption layer (per time unit and per one cluster) on the temperature of the substrate for two atomic types and two cluster sizes. This figure shows that the temperature of decomposition of the cluster depends on the cluster size and is higher for CF_4.

This agrees well with experiment [30,31]. In this experiment it was established that the transition from a one-layer to a two-layer structure for C_2H_4 occurs at $T=75$ K (or 0.34 in relative units) and for CF_4 at temperature $T=57.6$ K (0.43 in relative units).

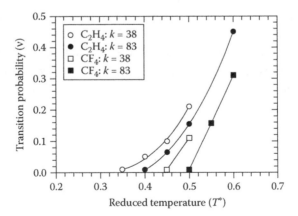

Figure 6.13   Dependence of probability of transition of atoms into the second layer on temperature.

Figure 6.13 shows that MD calculations predict the difference between two temperatures in the order of 0.1, very close to experimental data.

Figure 6.14 represents the dependence of the total energy at a constant temperature of $T^*=0.45$ for the adsorption atoms of C_2H_4. We can see that this dependence confirms experimental observation about the existence of a phase transition of the first kind: an island of two-dimensional gas adatoms on the surface.

The atomic size plays a significant role in this analysis and behavior. This is seen in Figure 6.15, where the results for CF_4 are depicted, for which the diameter s of the Lennard-Jones potential

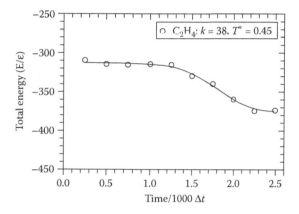

Figure 6.14 Time evolution of total energy of a C_2H_4 system with parameters $k=38$ and $T^*=0.45$.

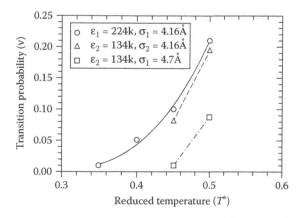

Figure 6.15 Dependence of probability of transition of adsorbed atoms on temperature for two parameter sets: $\varepsilon=134, 224$ K; $\sigma=4.16$ and 4.7 Å.

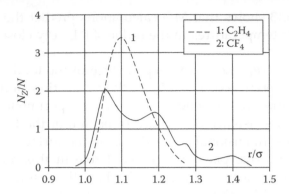

Figure 6.16   Radial distribution function of adsorption molecules in x,y plane for first layer of the film at $T^* = 0.45$.

was chosen equal to 4.16 Å, that is, equal to that of C_2H_4. In this case the results are almost similar for both species.

Figure 6.16 presents the radial distribution functions $g\ (r)$ of adatoms in the first layer. It can be seen that CF_4 adatoms are located on average at closer distances than for C_2H_4 which leads to a higher transition temperature.

Hence, the MD calculations are in good agreement with experiment. This obtaining can be used for predictive calculations by MD that are applicable for other atomic species on the surface.

6.9   Summary of Chapter 6

1. Brownian dynamics method was developed for calculating the cluster diffusion coefficient of Pt_n $(n = 2-6)$ clusters in the basic graphite surface. This dependence of the diffusion constant for large clusters D_n, where n is the number of atoms in it, can be approximated by a power law: $D_n = D_1\ n^{-0.87}$, which is close to experimental data.

2. The diffusion coefficient of Ptn clusters on a graphite basal surface were employed for the solution of microkinetic equations for the densities of clusters of sizes $n = 1-30$. The calculation results correctly describe the distribution of clusters on sizes, observed in experiments.

3. A direct calculation by MD of the cluster formation kinetics for deposition of Pt on graphite and Pd on MgO at room

temperature. The calculations were carried out for three model systems: two-dimensional isothermal, three-dimensional iso-thermal with iterations with the substrate atoms, and three-dimensional Brownian dynamics. It was shown that the latter two models correctly describe the experimental data for the case of weak adhesion.

4. The result of spinodal decomposition in a 2D supersaturated system agrees with the microkinetic equations and MD calculations.

5. MD calculations showed that the decomposition and decay of C_2H_4 and CF_4 clusters on a graphite surface at increasing surface temperature has character of phase transition of the first kind, which agrees with experiment. Calculated phase transition temperatures are in agreement with the existing experiments.

References

1. B. K. Vainshtein, *Modern Crystallography*, Springer: Berlin, 1981–c1988.

2. F. F. Abraham, Computational statistical mechanics: Methodology, applications and supercomputing, *Adv Phys* 35, 1–111, 1986.

3. R. A. Haefer, *Cryopumping: Theory and Practice*, Clarendon Press: London, 1989.

4. T. Takagi, *Ionized Cluster Beam Deposition and Epitaxy*, Noyes Publications: Park Ridge, NJ, 1988.

5. Yu. K. Tovbin, *Theory of Physical Chemistry Processes at a Gas–Solid Interface*, CRC Press: Boca Raton, FL, 1991.

6. U. Leuthäusser, Kinetic theory of adsorption and desorption, *Z Phys B* 44, 101–108, 1981.

7. K. Kinoshita, Mobility of small clusters on the substrate surface, *Thin Solid Film* 85, 223–238, 1981.

8. K. Takeuchi and K. Kinoshita, Mobility of gold clusters on amorphous carbon substrates I: Analysis of cluster density versus time relations. Part 1, *Thin Solid Film* 90, 27–30, 1981.

9. K. Takeuchi and K. Kinoshita, Mobility of gold clusters on amorphous carbon substrates II: Evaluation of surface diffusion coefficients. Part 2, *Thin Solid Film* 90, 31–35, 1981.

10. M. Harsdorff and G. Reiners, Mobility of small gold crystallites on the cleavage planes of alkali halides, *Thin Solid Film* 85, 267–273, 1981.

11. M. Volmer and A. Weber, Keimbildung in übersättigten Gebilden. *Z Phys Chem (Leipzig)* 119, 277, 1926.

12. J. A. Venables, C. D. T Spiller, and M. Hanbücken, Nucleation and growth of thin films, *Rep Prog Phys* 47, 399–459, 1984.

13. Z. Insepov and S. V. Zheludkov, Computer simulation of clustering on the surface, *18th Seminar of Middle-European Coop on Statist Physics*, Duisburg, Germany, 1991.

14. V. P. Zhdanov and K. I. Zamaraev, Lattice-gas model of chemisorption on metal surfaces, *Soviet Phys Usp* 29, 755, 1986.

15. C. E. Klots, Evaporation from small particles, *J Phys Chem* 92, 5864–5868, 1988.

16. A. Y. Valuev, Z. Insepov, and S. V. Zheludkov, Molecular-dynamic modeling of kinetics of cluster formation in supersaturated vapor (in Russian), *J Phys Chem* 61, 1109–1111, 1987.

17. A. Rahman, Correlations in the motion of atoms in liquid argon, *Phys Rev* 136, A405, 1964.

18. Z. Insepov and S. V. Zheludkov, Molecular kinetics of cluster formation in the dense fluids, *Z Phys D* 20, 453–455, 1991.

19. D. Kaschiev, *Nucleation*, Butterworth-Heinemann, 2000.

20. J. B. Adams and W. Hitchon, Solution of master and Fokker-Planck equations by propagator methods, applied to Au/NaCl thin film nucleation, *J Comp Phys* 76, 159–175, 1988.

21. J. C. Tully, Dynamics of gas-solid interaction: 3D generalized Langevin model used for fcc and bcc surfaces, *J Chem Phys* 73, 1975–1986, 1980.

22. A. Y. Valuev, Z. Insepov, S. V. Zheludkov, and Yu.V. Podlipchuk, Modeling the kinetics of cluster formation in a 2-component medium by the molecular-dynamics technique (in Russian), *J Phys Chem* 63, 1469–1475, 1989.

23. J. F. Hamilton, D. R. Preuss, and G. R. Arai, Nucleation and growth of vacuum deposition metal aggregates studied by electron microscopy, *Surf Sci* 106, 146–151, 1981.

24. Y. W. Lee and J. M. Rigsbee, The effect of dimer mobility on thin film nucleation kinetics, *Surf Sci* 173, 49–64, 1986.

25. W. A. Steel, The physical interaction of gases with crystalline solid, *Surf Sci* 36(1), 317–352, 1973.

26. S. M. Levine and S. H. Garofalini, Molecular dynamics simulation of Pt on a vitreous silica surface, *Surf Sci* 163, 59, 1985.

27. C. Kittel, *Introduction to the Solid State Physics*, 6th edn, New York: Wiley, 1986.

28. C. Chapon, C. R. Henri, and A. Cheman, Formation and characterization of small Pd particles deposited on MgO as model catalyst, *Surf Sci* 162, 747–754, 1985.

29. R. C. Reid and J. M. Prausnitz, *The Properties of Gases and Liquids*, New York: McGraw-Hill, 1977.

30. S. G. J. Mochrie, M. Sutton, R. J. Birgeneau, D. E. Moncton, and P. M. Horn, Multilayer adsorption of ethylene on graphite: Layering, prewetting and wetting, *Phys Rev B* 1, 263–273, 1984.

31. Q. M. Zhang, H. K. Kim, and M. H. W. Chan, Nonwetting growth and cluster formation of CF4 on graphite, *Phys Rev B* 34, 2055, 1986.

32. M. Bienfait, Wetting and multilayer adsorption, *Surf Sci* 162, 411–420, 1985.

33. M. Drir, H. S. Nham, and G. B. Hess, Multilayer adsorption and wetting: Ethylene on graphite, *Phys Rev B* 33(7), 5145, 1986.

34. J. I. Seguin, J. Suzanne, M. Bienfait, J. G. Dash, ' and J. A. Venables, Complete and incomplete wetting in multilayer adsorption: High-energy electron-diffraction studies of Xe, Ar, N_2, and Ne films on graphite, *Phys Rev Lett* 51, 122–125, 1983.

CHAPTER 7

Multiscale Concept for Condensation in Rarefied Gases

7.1 Introduction

One of the most important physical quantities of condensing vapor that can be measured in experiments is the cluster size distribution function. To determine this function for condensations in supersonic flows expanding through nozzles or during condensation of metal vapors behind shock wave fronts, one usually uses the Classical Nucleation Theory (CNT) [1] or kinetic equations for the size distribution function, mostly according to Szilard's scheme [2–8].

However, the deviation between theory and experimental data used to be so large that this inconsistency between theory and the measurement of condensation was given the name the Lothe–Pound paradox [9].

One of the sources of uncertainties in the theory based on kinetic equations could have been the uncertainty related to the gas-kinetic approach with which the kinetic coefficients of cluster growth and decay were calculated.

This fact was not surprising, since a rigorous calculation of the cluster kinetic coefficients assumes the solution of simultaneous interactions of many bodies. Therefore (see, e.g., [1] and [6]), a reverse problem was solved instead; namely, the probabilities of the cluster growth/decay were restored using adjustment between theory and experiment.

In this chapter, a new task was developed and solved on a systematic determination of the kinetic coefficients of cluster growth and decay using molecular dynamics. To compare theory with experiment, the rate constants calculated in this work were applied to explain the condensation process of vaporized iron atoms in a supersonic flow of Ar and to understand the kinetics of condensation of iron atoms behind the shock wave front.

It was shown that this way that obtaining a quantitative agreement between the theory and the experimental data [10,11].

7.2   Calculation of Growth and Decay Constant Rates

In Chapter 2, a molecular dynamics (MD) model of cluster size evolution was developed that calculates the probabilities of various outcomes after an energetic collision of a single atom (a monomer) with the cluster placed in the origin of the coordinate system. In the present section, the MD model of cluster size evolution was applied for calculating the rate constants of the reactions between the gas atom and a cluster leading to association of a single atom or a dissociation of atoms from the existing cluster or molecule. The smallest clusters consist of two or three atoms bound together via week van der Waals forces. They are called dimers and trimers, respectively.

Clusters were placed into the origin of the Cartesian coordinate system. A gas atom (monomer) approached the cluster along the z-axis with a velocity v_z (see Figure 2.1) and having an impact parameter p.

The following outcome events were observed as a result of the simulations: association of the projectile atom with the cluster of n atoms (i.e., formation of a bigger cluster consisting of $n+1$ atoms); transfer of a fraction of kinetic energy of the monomer to the cluster and bouncing back of the monomer with a lower kinetic energy (an event of cluster excitation $An \rightarrow An^*$); and, finally, bouncing back of the monomer with an increased energy (a cluster stabilization event $An^* \rightarrow An$).

Figure 7.1 shows the relative numbers of clusters of size $n+1$, $N(t)/N_0$, formed in the collision of a monomer with a cluster containing n atoms. Here $N(t)$ is the number of clusters survived at time

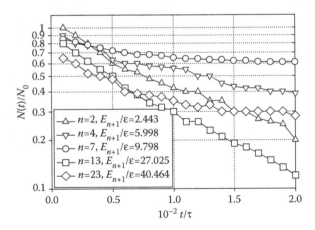

Figure 7.1   Number of clusters survived by time t, $\tau = \sigma \, (m/\varepsilon)^{1/2}$, time is given in Lennard-Jones units: $n=2$, $E_{n+1}/\varepsilon = 2.443$; $n=4$, $E_{n+1}/\varepsilon = 5.998$; $n=7$, $E_{n+1}/\varepsilon = 9.798$; $n=13$, $E_{n+1}/\varepsilon = 27.025$; $n=23$, $E_{n+1}/\varepsilon = 40.464$.

t, N_0 is the total number of clusters initially, at $t=0$. As is seen in the figure, except for the case $n=2$, these results show two characteristic times in the time dependence of the kinetics of excited cluster decay. The first characteristic time is larger than 10^{-14} s and it is clearly visible for the collision of a monomer with a cluster of 23 atoms. This characteristic time is of the same order of magnitude as the collision time $\tau \sim \sigma/v$ and reflects the nonequilibrium character of the decay process of a collisional complex in gas phase.

The second slope in Figure 7.1 corresponds to an equilibrium cluster decay of a single atom evaporating from the cluster's surface. This slope can be described by the following dependence [12]:

$$\ln\left(\frac{N(t)}{N_0}\right) = -k_d\left(t - t_0\right),\tag{7.1}$$

where:

k_d is the evaporation rate constant (s^{-1})

t_0 is the threshold time, set as an average time of redistribution of an excess energy of monomer over the internal degrees of freedom of a cluster

The evaporation constants obtained in these calculations were well fitted by a dependence given by the statistical theory of reactions RRK (see Figure 7.2) [13]:

$$k_d = v\left(1 - \frac{E_0}{E_{n+1}}\right)^{s-1},\tag{7.2}$$

where:

E_0 is the activation threshold for evaporation at $T=0$

E_{n+1} is the total internal energy of the cluster containing $n+1$ atoms

v and s are the parameters of the model.

The energy levels in Equation 7.2 are counted from the global minimum of energy of the cluster in Equation 7.2. Table 7.1 contains the values of v and s obtained by evaluating k_d using the least square method.

The rate coefficients of cluster formation can be calculated from Figure 7.1 by a linearized extension of the equilibrium dependence of k_d on time to the y-axis [12]. Figure 7.3 represents the dependence of the cluster formation coefficient α_f on the impact parameter

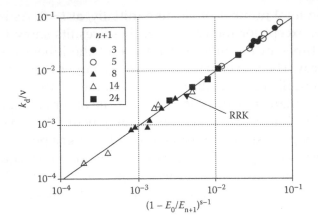

Figure 7.2 Parameters of the RRK theory calculated by MD. The straight line corresponds to the statistical theory of reactions RRK. (From K. A. Holbrook, et al., *Unimolecular Reactions*, 2nd edn, Wiley: New York, 1996.)

Table 7.1 Parameters v and s in Equation 7.2 Calculated by MD in This Work

$N+1$	$-U_0/\varepsilon$	E_0/ε	s	v, s^{-1}
3	3.000	2.000	2.93	1.02×10^{12}
5	9.104	3.104	5.60	0.52×10^{12}
8	19.821	4.228	13.70	5.05×10^{12}
14	47.845	3.518	55.40	4.91×10^{12}
24	96.504	3.660	58.50	1.94×10^{12}

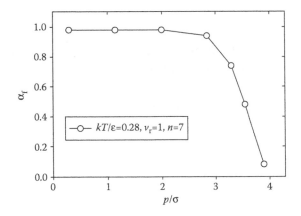

Figure 7.3 Dependence of the cluster formation coefficient on the impact parameter $p^* = p/\sigma$ for $n=7$ and $T^* = kT/\varepsilon = 0.28$. From this figure it is seen the existence of the maximum impact parameter $p^*_m \approx 4$.

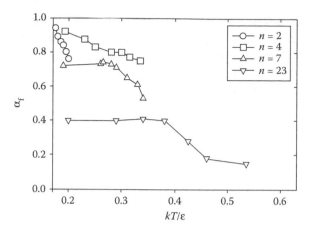

Figure 7.4   Dependence of the cluster formation coefficient α_f on cluster temperatures for $n=2$–23.

$p^*=p/\sigma$ for $n=7$ and $T^*=kT/\varepsilon=0.28$. This figure shows that there is a maximum impact parameter of $p^*_m\approx4$.

Figure 7.4 shows similar dependences for other cluster temperatures.

For modeling collisions between a monomer A and a dimer A_2, the impact dissociation of the dimer was found as having the highest probability; that is, the following reaction was the most probable:

$$A_1 + A_2 \overset{\alpha_{sd}}{\rightarrow} A_1 + A_1 + A_1.$$

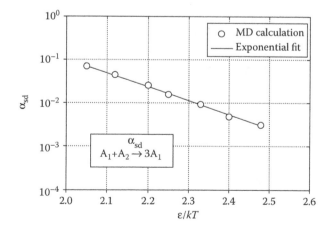

Figure 7.5   Dependence of formation rate constant α_{sd} on the reciprocal cluster temperature.

The dependence of α_{sd} on cluster temperature is shown in Figure 7.5. The following processes occurring at the collision of monomer with cluster were the processes of excitation (with a probability α_e) and stabilization (with a probability α_s).

These probabilities were determined by the expression [14]

$$\alpha_i = \frac{1}{\left(1 + k_B T \big/ \left(\langle \Delta E_i^2 \rangle\right)^{1/2}\right)} (i = e, s),$$ (7.3)

Figures 7.6 and 7.7 represent the dependences of $\langle \Delta E_e^* \rangle$ and $\langle \Delta E_s^* \rangle$ on temperature for clusters with $n = 2, 4, 7$, and 13.

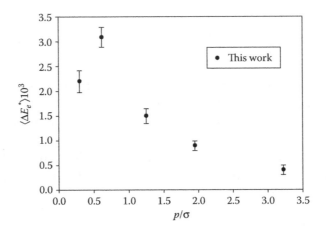

Figure 7.6   Dependences of $\langle \Delta E_e^* \rangle$ on the impact parameter.

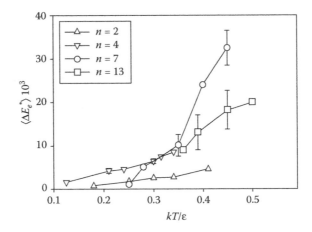

Figures 7.7   Dependences of $\langle \Delta E_e^* \rangle$ on temperature for clusters with $n = 2$, 4, 7, and 13.

The methodology of calculating the probability of collision events during collisions of monomers and clusters used in this section can be easily adopted for the collisions of atoms of different types.

The following processes were modeled:

$$Ar + Fe_n \rightarrow Fe_n^* + Ar,$$

$$Fe' + FeCo \rightarrow Fe' + FeCo,$$

$$Fe' + FeCo \rightarrow Fe'Co + Fe, \qquad (7.4)$$

$$Fe + FeCo \rightarrow Fe + Fe + Co,$$

$$Fe + FeCo \rightarrow Fe_2 + Co.$$

These reactions correspond, respectively, to the processes of Fe_n cluster excitation, elastic scattering of Fe atoms on a $FeCo$ molecules, and dimer formation. The Lennard-Jones parameters for interaction of atoms Ar and Fe were obtained by the Lorentz–Berthelot rule:

$$\varepsilon_{Ar-Fe} = (\varepsilon_{Ar} \cdot \varepsilon_{Fe})^{1/2}, \sigma_{Ar-Fe} = \frac{(\sigma_{Ar} + \sigma_{Fe})}{2},$$

where the following data were used: $\varepsilon_{Ar} = 1.656 \times 10^{-14}$ erg, $\sigma_{Ar} = 3.405$ Å [15], $\varepsilon_{Fe} = 1.064 \times 10^{-12}$ erg, $\sigma_{Fe} = 2.95$ Å [16].

Figures 7.6 through 7.11 show the dependences of the average energy transfer $\langle \Delta E_i^* \rangle$ and α_i ($i = e, s$) for clusters Fe_n ($n = 2, 13, 23$).

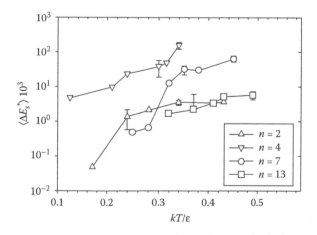

Figure 7.8   Dependences of $\langle \Delta E_s^* \rangle$ on temperature for clusters with $n = 2$, 4, 7, and 13.

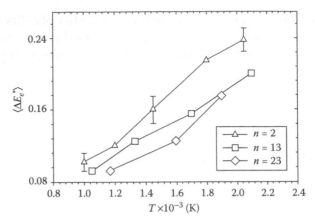

Figure 7.9   Dependences of the average excitation energy transfer $\left\langle \Delta E_e^* \right\rangle$ for clusters Fe_n ($n=2, 13, 23$).

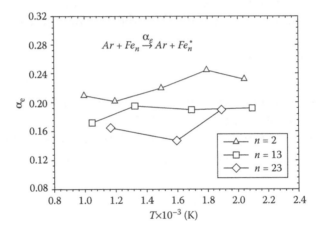

Figure 7.10   Dependences of the excitation rate coefficient α_e for clusters Fe_n ($n=2,13,23$) due to collision with Ar atom.

Figure 7.12 presents the dependence of the stabilization rate constant for iron clusters Fe_n ($n=2$, 13, 23). The dependence of the dimerization probability of iron atoms on the reciprocal temperature is given in Figure 7.13.

Figures 7.14 and 7.15 show the dependences of the sticking coefficients of a monomer to a cluster for two values of cluster temperature and monomer concentrations. The methodology of calculations of the sticking coefficients was discussed in Section 2.2, and the quantities ν, s, E_{n+1}, and k_s were calculated by MD. The calculations

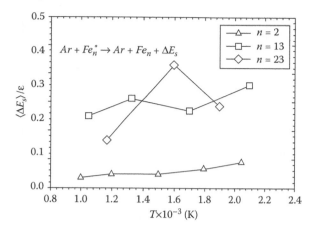

Figure 7.11   Dependences of the stabilization energy transfer for clusters Fe_n ($n = 2, 13, 23$) due to collision with Ar atom.

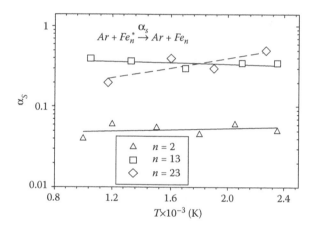

Figure 7.12   Dependences of the stabilization rate constant α_s for clusters Fe_n ($n = 2, 13, 23$).

show a monotonous increase of the sticking coefficient with increase of n. Figures 7.14 and 7.15 use a variable $n^{-2/3}$ since such a choice of variable allows better fitting of the sticking coefficient in an area on large n.

In Figures 7.14 and 7.15, the calculated data in this work are compared to MD calculations carried out by Bedanov [17] and the data extracted from the experiments by Lu and Yang [6].

As can be seen from the figures, the agreement is satisfactory, if we take into account that the methodology [17] corresponds rather to the limit of high pressure, and the evaluation of the experiments

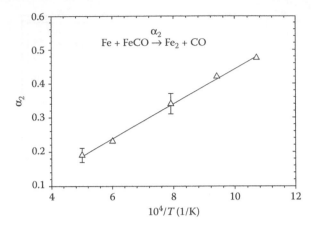

Figure 7.13   Dependence of the dimerization probability on the reciprocal temperature.

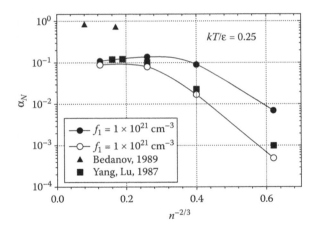

Figure 7.14   The data calculated in this work are compared to the MD calculations carried out by Bedanov and the data extracted from the experiments by Lu and Yang for $kT/\varepsilon = 0.25$. (Data from S.-N. Yang and T.-M. Lu, *Solid State Commun*, 61, 351–354, 1987; V. M. Bedanov, *Chem Phys*, 8, 117–121, 1989.)

in [6] was conducted on the base of the classical liquid droplet theory.

In Table 7.2 the calculated values of the average fraction of energy transferred to Fe_n in an act of collision with an Ar atom are compared to experiments [11,18].

As one can see, our data for $n = 2$–23 are satisfactory as a whole, if we take into account that the experimental data are given as

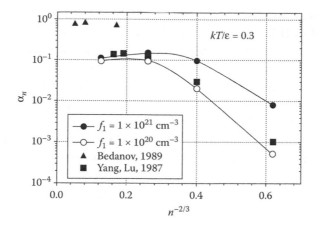

Figure 7.15   The data calculated in this work are compared to the MD calculations carried out by Bedanov and the data extracted from the experiments by Lu and Yang [6] for $kT/\varepsilon = 0.3$. (Data from S.-N. Yang and T.-M. Lu, *Solid State Commun*, 61, 351–354, 1987; From V. M. Bedanov, *Chem Phys*, 8, 117–121, 1989.)

Table 7.2   Comparison of MD Calculations with Experiment

$\langle \Delta E_s \rangle / \varepsilon$	MD (this work) $T = 1000–2100\,K$	Experiment [18] $T = 1609–1814\,K$	Experiment [11] $T = 1340–1980\,K$
2	0.03–0.07	0.04	–
13	0.21–0.29	0.14	–
23	0.15–0.36	0.18	–
1000	0.15–0.36	0.38	0.05–0.13

Source: S. K. Aizatullin, et al., *Chem Phys*, 46, 851–856, 1985; A. J. Freund and S. H. Bauer, *J Chem Phys*, 81, 994–1000, 1977.

an average for the whole temperature range. Extrapolation of our results into the area $n \sim 1000$ corresponds well to experiments.

7.3   Calculation of Evaporation Rate Constant at Low Temperatures

As was mentioned in Section 2.2, the MD method allows one to calculate the evaporation constant rate k_d of clusters at intermediate temperatures. As the temperature declines, the efficiency of the MD method significantly worsens due to the enormous increase in computation time.

This section discusses the calculation results of the method developed in this work (Section 2.5) representing a combination of the theory of transition state (TST) and the MD method [19,20].

A 14-atomic cluster where atoms were interacting via a Lennard-Jones potential was studied. Thirteen atoms were placed in icosahedral positions, while the last (14th) atom was placed on the top of the almost spherical 13-atomic cluster.

To calculate the evaporation constant rate, a compensation potential method was adopted that can reduce the complicated task of computing the rate constant in the system of N interacting particles to the dynamics of a single particle moving in a shallow, relative to the system's temperature, potential W.

Figure 7.16 shows the real potential V, the compensating potential $U(s)$, and the difference between these two potentials W. It is clearly seen in this figure that the real potential V is too deep for this temperature $kT/\varepsilon = 0.24$. However, the difference W is of the same order of magnitude as the kinetic energy of the evaporating particle.

The canonical ensemble was built as follows (see Figure 2.2). A cluster was surrounded by an ideally reflecting sphere of radius S_0. The modulus of the reflected particle from the sphere was selected from a Maxwellian distribution at T. A function $g(s)$ and then k_{TST}

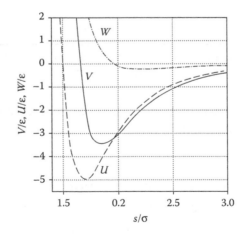

Figure 7.16   Real potential V, compensating potential $U(s)$, and the difference potential $W = V - U$. It is clearly seen that V is a deep potential at $kT/\varepsilon = 0.24$, but W has a depth comparable with the kinetic energy of the evaporated atom.

Equation 2.26 was calculated. A dynamic factor F was obtained according to Tully's methodology [21]:

$$F = \alpha \exp\left[-\frac{V(s_0)}{k_B T}\right],\qquad(7.5)$$

where α is the sticking coefficient of an atom to the cluster.

To verify the method, the distribution function of the cluster atoms was calculated and compared to a Maxwellian one (see Figures 7.17 and 7.18).

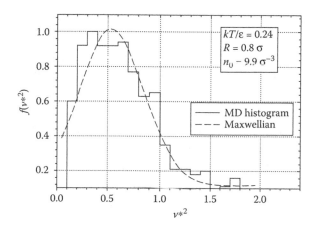

Figure 7.17   Velocity square distribution function of the cluster atoms calculated and compared to a Maxwellian.

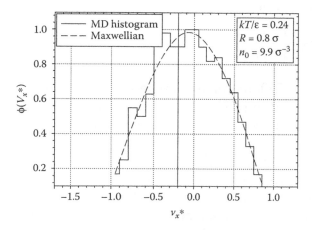

Figure 7.18   Velocity component distribution function of the cluster atoms calculated and compared to a Maxwellian.

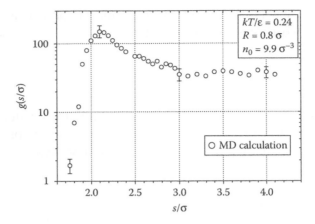

Figure 7.19   Function $g(s)$ calculated in this work. It was obtained that the value $S_0 = 4\sigma$ guarantees the necessary level of accuracy and the computational expense of the model.

Figure 7.19 represents the function $g(s)$ calculated by this method. It was obtained that the value $S_0 = 4\sigma$ guarantees the necessary level of accuracy and the computational expense of the model.

The coefficient α was calculated by averaging the results over as many as 1000 trajectories and the result $\alpha = 0.8$ was found for the temperature region $kT/\varepsilon = 0.16–0.35$.

Figure 7.20 presents the results of the calculations of the rate constant of evaporation in relative units: $k_d^* = k_d(m\sigma^2/\varepsilon)^{1/2}$ and $T^* = kT/\varepsilon$. The results obtained in this calculation can be well fitted by an Arrhenius law (line 1):

$$k_d = \nu \exp\left[-\frac{E_d}{kT}\right]. \tag{7.6}$$

In Figure 7.20, the high-temperature results of the MD calculations obtained in [17] are also shown for comparison (line 2). In Table 7.3, the values of the pre-exponential ν and the energy of activation E_d obtained in our calculations and data from [17] for clusters of Ar and Fe are presented.

As it follows from comparing our calculations with the high-temperature calculation data [17], there is a jump (step) in the dependence k_d $(1/T)$ at a temperature interval of $kT/\varepsilon = 0.3–0.4$. This fact can be explained by cluster melting and the following loosening of the atomic structure. It is remarkable that the calculation of

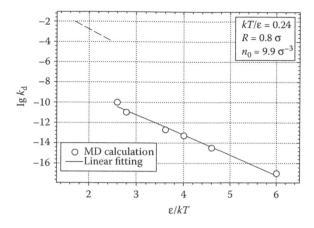

Figure 7.20   The result of calculations of the rate constant of evaporation in relatives units: $k_d^* = k_d(m\sigma^2/\varepsilon)^{1/2}$ and $T^* = kT/\varepsilon$. The results obtained in this calculation can be well fitted by an Arrhenius law $k_d = \nu \exp[-E_d/kT]$ (shown by line).

Table 7.3   Comparison of This Work with Direct MD Simulations

Substance	Ar		Fe	
Reference				
Parameters	This Work	[17]	This Work	[17]
ν, s^{-1}	9.312×10^6	1.875×10^{13}	7.267×10^7	1.464×10^{14}
E_d, erg	8.644×10^{-14}	8.755×10^{-14}	5.554×10^{-12}	5.625×10^{-12}

Source: V. M. Bedanov, *Chem Phys*, 8, 117–121, 1989.

a 13-atomic cluster melting by Monte Carlo method obtained the melting temperature $kT/\varepsilon = 0.29$ [22].

Therefore, the melting temperature of a cluster of 14 atoms must be larger than 0.29. If this is the case, the large difference in the pre-exponential obtained in our calculations and in [17] can be ascribed to the significant growth of activation entropy at the melting and loosening of the transition complex [13].

7.4   Summary of Chapter 7

1. Molecular dynamics calculations of the kinetic rate constants and decay of clusters in rarefied gases were conducted in this chapter. The dependences on temperature, energy,

and cluster size for the probabilities of formation, stabilizations, and evaporation of clusters were obtained.

2. The equilibrium coefficient of cluster evaporation was calculated using the statistical theory of Rice-Ramsperger-Kassel (RRK). The probabilities of the iron dimer formation were calculated in the presence of molecules of FeCO.

3. The evaporation rate constant of a 14-atomic cluster was calculated at a low temperature $kT/\varepsilon = 0.38–0.17$ where the traditional MD methods are inapplicable.

4. A jump (step) in the evaporation rate constant was found that can be explained by cluster melting and the following structure loosening.

References

1. Y. I. Petrov, *Clusters and Small Particles*, Nauka: Moscow, 1986.

2. Y. G. Korobeinikov and Y. S. Kusner, Calculation of the dimerization rate constant for atoms, *Tech Phys Lett* 50, 1581–1582, 1980.

3. Y. G. Korobeinikov, Calculation of the rate constant for three-body recombination, *Phys Combust Explos* 11(6), 863–870, 1975.

4. A. A. Vostrikov, Y. S. Kusner, A. K. Rebrov, and B. G. Semyachkin, Formation of dimers in molecular gases, *Tech Phys Lett* 3(24), 1319–1323, 1977.

5. D. J. Frurip and S. H. Bauer, Homogeneous nucleation in metal vapors. 4. Cluster growth rates from light scattering, *J Phys Chem* 81(10): 1007–1015, 1977.

6. S.-N. Yang and T.-M. Lu, The sticking coefficient of Ar on small Ar clusters, *Solid State Commun* 61(6): 351–354, 1987.

7. W. G. Dorfield and J. S. Hudson, Condensation in CO_2 free jet expansions. 2. Growth of small clusters. *J Chem Phys* 59(3): 1261–1265, 1973.

8. D. L. Bunker, Mechanics of atomic recombination reactions, *J Chem Phys* 32(4): 1001–1005, 1960.

9. J. Lothe and G. M. Pound, Reconsideration of nucleation theory, *J Chem Phys* 36(8): 2080–2085, 1962.

10. T. A. Milne and F. T. Greene, General behavior and equilibrium dimer concentration, *J Chem Phys* 47(10): 4095–4101, 1967

11. S. K. Aizatullin, I. S. Zaslonko, V. N. Smirnov, and A. G. Sutugin, Study on condensation of iron vapor at decay in shock waves (in Russian), *Chem Phys* 46(6): 851–856, 1985.

12. J. W. Brady, J. D. Doll, and D. L. Thomson, Cluster dynamics: A classical trajectory study of $A+A_n \rightarrow A^*_{n+1}$, *J Chem Phys* 71(6): 2467–2472, 1979.

13. K. A. Holbrook, M. J. Pilling, and S. H. Robertson, *Unimolecular Reactions*, 2nd edn, Wiley: New York, 1996.

14. V. N. Kondratiev and E. E. Nikitin, *Gas-Phase Reactions: Kinetics and Mechanisms*, Springer: New York, 2011.

15. J. O. Hirschfelder, C. F. Curtiss, and R. B. Bird, *Molecular Theory of Gases and Liquids*, p. 1249, Wiley: New York, 1954.

16. S. H. Bauer and D. J. Frurip, Homogeneous nucleation in metal vapors. 5. A self-consistent kinetic model, *J Phys Chem* 81(10) 1017–1024, 1977.

17. V. M. Bedanov, Condensation coefficient of small clusters and its effect on the nucleation rate. Calculation by MD method (in Russian). *Chem Phys* 8(1): 117–121, 1989.

18. A. J. Freund and S. H. Bauer, Homogeneous nucleation in metal vapors. 2. Dependence of the heat of condensation on cluster size. *J Chem Phys* 81(10): 994–1000, 1977.

19. Z. Insepov and E. Karataev, Molecular dynamics simulation of infrequent events: Evaporation from cold metallic clusters, *Phys Chem Finite System: From Cluster Crystal* 1: 423–427, 1992.

20. Z. Insepov and E. Karataev, Molecular dynamics calculation of the evaporation rate of cold clusters (in Russian), *Tech Phys Lett* 17(24): 36–40, 1991.

21. E. K. Grimmelman, J. C. Tully, and E. Helfand, Molecular dynamics of infrequent events: Thermal desorption of xenon from a platinum surface, *J Chem Phys* 74(9): 5300–5310, 1981.

22. R. D. Etters and J. Kaelberer, Thermodynamic properties of small aggregates of rare-gas atoms, *Phys Rev A* 11(3): 1068–1079, 1975.

CHAPTER 8

Nucleation and Condensation in Gases

This chapter gives a brief review of the Classical Nucleation Theory (CNT) in gases. The CNT gives good predictions for cloud chambers where the size of the critical nucleus is rather large. However, it is not applicable to systems at very high supersaturation degrees such as expansion of gases in supersonic nozzles where the critical nucleus size becomes less than unity. Atomistic theories, such as molecular dynamics, can significantly improve the applicability of the CNT to such systems and improve comparison with experiment. In this chapter, the CNT was modified by adding the kinetic rate coefficients calculated by MD for the high supersaturation case of condensation of metal vapor behind the shock wave front.

8.1 Introduction

The basic understanding of the nucleation theory was achieved almost 100 years ago by Gibbs [1]. His works published in 1904 have built the foundation of the so-called Classical Nucleation Theory (CNT). Further development of the CNT have been provided by Volmer and Weber [2–5], Farkas [6], Becker and Döring [7], Zel'dovich [8], Frenkel [9], Lothe and Pound [10,11], and Reiss et al. [12]. The CNT was capable of predicting the earlier experimental results obtained by Volmer and Flood [4].

The CNT consists of two entirely different approaches that were developed separately and are not always related to each other:

1. Thermodynamics of nucleation, which was mostly developed by Gibbs in his classical work [1] and by Volmer and Weber [2,3,5]. The main parameters of the thermodynamic theory are the critical radius r^* of the droplet and the nucleation rate, J (cm^{-3} s^{-1}), which is the number of nuclei formed per unit volume, per unit time.

2. Kinetic theory of condensation, which is based on the absolute rate theory of chemical reactions and consists of rate equations obtained using the mean-field theory. These kinetic equations are solved for concentrations of the new phase nucleus of different sizes (radii or number of atoms) and the nucleation rate.

In the following we briefly discuss both approaches.

8.2   Thermodynamics of Nucleation

Volmer and Weber introduced key parameters of the condensation process in a supersaturated vapor [2]. According to these authors, the level of supersaturation in a condensing system can be characterized by a certain critical size of a droplet of the new phase. Thus the gas will be supersaturated against the droplets whose radii are larger than that of the critical droplet. As for the smaller droplets, the gas will be undersaturated relative to the smaller sizes, thus these droplets will be in equilibrium with the vapor. The frequency of such processes is defined by Boltzmann's relation between the entropy S of formation of a droplet with the radius r_n from vapor at pressure P, at constant volume and energies. The probability of finding such a droplet in the system will be proportional to $\exp(-S/k)$. If the number of molecules in the system is much larger than the number of molecules in the droplet, n, this entropy time absolute temperature, T, is equal to isothermal and reversible work W needed to form such a droplet by an external system. Volmer [2,3,5] proposed the following processes that can lead to the formation of a droplet from the vapor in an isothermal and reversible process:

1. Extraction of n molecules from the vapor

2. Expansion of the droplet from P_n to $P\infty$

3. Condensation on a flat liquid surface

4. Formation of a droplet from the liquid

The sum of these steps should give the required work W. Steps 1–3 are canceled by each other. The rest of the work gives

$$W = -nkT \ln \frac{P_n}{P_\infty} + \sigma A. \tag{8.1}$$

Here, the first term is equal to the negative work of formation of a bulk liquid with the droplet's volume and the second term is equal to the formation work of the droplet's surface A.

Using Thompson's (Lord Kelvin's) formula for the pressure P_n above the droplet built by n molecules having a surface A gives the following relations:

$$n_0 = \frac{4\pi}{3} r_n^3 n_0, \tag{8.2}$$

$$A = 4\pi r_n^2, \tag{8.3}$$

$$\ln \frac{P_n}{P_\infty} = \frac{2\sigma}{kTn_0} \frac{1}{r_n}. \tag{8.4}$$

The ratio P_n/P_∞ is referred to as the supersaturation S. The droplet formation work W_n is the reversible work needed for producing the droplet containing n molecules:

$$W = \frac{1}{3}\sigma A. \tag{8.5}$$

Therefore, the main result of the VW theory can be written as the following equation for the nucleation rate:

$$J = K_v e - \frac{\sigma A_n}{3kT}, \tag{8.6}$$

where:

   σ is the surface tension
   A_n is the surface of the critical embryos of the new phase at pressure P_n
   J is the stationary rate of homogeneous nucleation (s^{-1} cm^{-3})

The pre-exponential coefficient K_V cannot be predicted by thermodynamics and should be obtained in a different way. Farkas obtained it using a gas-kinetic theory [6].

The derivation of the kinetic factor K_V given by Farkas [6] and summarized by Becker and Döring in [7] will be presented below. Let's consider a quasi-stationary condensation process where the following conditions are fulfilled: the very large volume of the condensation occurs in vapor at constant pressure P and temperature T which is provided by a continuous supply of single molecules.

To keep the process in a stationary state, the droplets with certain sizes s larger than the critical size n^* should be removed from the system and counted. Obviously, this definition gives the same stationary homogeneous nucleation rate introduced by Volmer et al. [2,3,5]. Under this consideration, the vapor will contain a stationary distribution of droplets with different sizes that were obtained in [4,7–9]. Let Z_v be the number of droplets containing v molecules, Z_1 the number of single molecules provided by a constant supply of gas from external source, and Z_s the total number of droplets of the critical size, which is set to zero: $Z_s = 0$.

The flux in the space of the droplets' sizes J can be represented as a sum of two fluxes: one flux is defined by the evaporation of molecules from the droplets' surface and another one by an opposite process of association of single molecules into the droplet from the vapor.

If $q_v\, dt$ is the probability of evaporation and $\alpha_0 dt$ is the probability of association, then the total flux can be written as

$$J = a_0 Z'_{v-1} - q_v Z'_v. \tag{8.7}$$

Introducing $\beta_v = \alpha_0 / q_v$, we can obtain

$$Z'_v = Z'_v \beta_v - \frac{J}{\alpha_0} \beta_v, \tag{8.8}$$

where $\beta_v = 1$ for $v = n^*$ and

$$\beta_v = \frac{p_n}{p_v} = e^{2\sigma/\rho_0 kT(1/r_n - 1/r_v)}, \tag{8.9}$$

The coefficient (8.9) is a monotonic function of v for $v < n$, and after several simplifications, using an electrical circuit analogy, Becker and Döring obtained the following expression (see [6], Equation 8.13):

$$J = \frac{a_0 Z'_1}{n} \sqrt{\frac{A'}{3\pi}} e - \frac{\sigma A_n}{3kT}. \tag{8.10}$$

Farkas earlier obtained an exactly identical expression [4].

For a practical comparison of Equation 8.8 for the condensation nucleation rate with the experimental data obtained by Volmer and Flood [4] for water vapor at $T = 260°$ and $275°$, Equation 8.13 can be rewritten as follows:

$$\ln J = 49 + \ln \frac{p_\infty^2 \rho}{\lambda M^{3/2} M \sigma^{3/2}} + 2x + 2\ln x - 17.7 \cdot \left(\frac{\sigma}{T}\right)^3 \left(\frac{M}{T}\right)^3 \cdot \frac{1}{x^2}.$$

(8.11)

8.3 Kinetics of Nucleation

The main characteristic of the kinetic rate theory is the cluster size distribution function $c(l)$ or c_l. The kinetic rate theory approach is called the Szilard-Farkas-Becker-Döring theory [3–9,13–18].

According to the theory proposed by Becker and Döring (BD) in 1935 [7], a supersaturated two-component system consists of small atomic (molecular) aggregates (or clusters) of various sizes $n = \{1 - N\}$, where the clusters of size 1are called monomers.

Some limitations of the BD theory include the following mechanisms that are responsible for changing the cluster sizes in time:

1. Only monomers can diffuse and associate with larger clusters.

2. Dissociation from a bigger cluster A_i, containing n atoms, occurs via emission of monomers A_1:

$$A_n + A_1 \leftrightarrows A_{n+1},$$

(8.12)

where the rate of processes that lead to the cluster's growth are denoted J_n and the rate of opposite processes of cluster size decrease are denoted J_{n-1}. If J_n is positive, the clusters tend to grow; if negative, the sizes decrease, and J is defined as the number of times that happens per unit volume per unit time.

The total amount of clusters per unit volume, the concentration of n-clusters c_n, can be obtained from the following balance Equation 8.12:

$$\frac{dc_n}{dt} = J_{n-1} - J_n \, (n \geq 2).$$

(8.13)

For the specific case of monomer concentration, the equation will be different than Equation 8.13 since a cluster of 2 atoms can only be created from two monomers* and monomers are involved in association to any cluster size:

$$\frac{dc_1}{dt} = -2J_1 - \sum_{n=2}^{\infty} J_n.$$

(8.14)

---

* This is not true for dimer formation in a rarefied system.

The system of kinetic Equations 8.13 and 8.14 needs a constitutive relation between the concentrations, and the rate J_n is the net flux from the size $n - 1$ to n:

$$J_n = \alpha_n c_1 c_n - b_{n+1} c_{n+1} \, (n \geq 1),$$
(8.15)

where the coefficients α_n and b_{n+1} are independent of time, where α_n is the kinetic rate coefficient for association of a monomer with an n cluster and b_{n+1} is the kinetic coefficient related to a thermal desorption rate of an $n + 1$ cluster where it emits a monomer.

The system Equation 8.13 is called the Becker–Döring kinetic equation. Burton [13] added Equation 8.14 to make the system conservative, that is, the density $\rho = \sum_{n=1}^{\infty} n c_n = \text{const.}$

The Becker–Döring theory makes an assumption that only monomers may interact with clusters.

The equilibrium solution of the Equations 8.13 and 8.14 leads to a Boltzmann-type formula:

$$c^* = c_1 \exp\left(\frac{-\Delta G^*}{kT}\right).$$
(8.16)

Here, ΔG^* is the free energy of formation of the critical nucleus:

$$\Delta G^* = \frac{16\pi\sigma^3}{3\Delta G_v^2},$$
(8.17)

where:

σ is the surface tension

ΔG_v is the free energy change per unit volume or, in terms of the supersaturation ratio, $\ln (p/p_0)$:

$$\Delta G^* = -\left(\frac{kT}{\Omega}\right)\ln\left(\frac{p}{p_0}\right).$$
(8.18)

The rate of nucleation (8.15) was calculated by Becker and Döring and Zel'dovich [7,8] as follows:

$$J = \frac{\gamma\left(4\pi r^{*2}\right)p}{(2\pi mkT)^{1/2}} c_1 \exp\frac{\left(-\Delta G^{*\prime}\right)}{kT},$$
(8.19)

where γ is the Zel'dovich factor that corrects the departure of the steady state from an equilibrium concentration of critical nuclei and the critical radius r^* is given by

$$r^* = \frac{-2\sigma}{\Delta G_v}. \tag{8.20}$$

Since the pre-exponential factor in Equation 8.19 is of the order of 10^{25}, the Zel'dovich factor $\gamma \approx 0.01$ is relatively insignificant. The dependence (Equation 8.19) on supersaturation is so steep that the critical supersaturation is not significant. Therefore, there is a rough agreement between (Equation 8.19) and the experimental data [12].

The above expressions are not exactly correct for very small critical sizes where the CNT becomes controversial. Lothe and Pound [10,11] showed that the CNT has inherent errors that can be corrected by adding the following contributions: quantum statistical contributions to the free energy of formation of the nuclei in the process of condensation. This contribution to the critical concentration of vapor was found by calculating the absolute entropy of the nucleation embryos and in the case of water condensation. For a nucleus containing 100 molecules, this contribution is approximately $\Delta G_s \approx 8kT$, which leads to a lower concentration of the critical nuclei by a factor of 10^{-3}. The two other contributions are the translational and rotational contributions: $-24.4\ kT$ for the translational and $-20.8\ kT$ for the rotational motions, which together with the quantum correction reduce the concentration by a factor of 10^{17}.

Therefore, Lothe and Pound showed that the Becker–Döring theory was basically inaccurate and comparison with experiments of Wilson and Powell [19] clearly showed a large discrepancy. This deviation between theory and experiment can be understood, if one assumes that instead of being at a stationary process, the condensing system experiences rather a transient state situation, where sticking of vapor atoms to small embryos were nonaccommodating, that is, the collisions of atoms with the smallest nuclei were incomplete.

The probabilities of sticking of vapor particles (atoms, molecules) are therefore important fundamental constants that cannot be obtained by the Classical Nucleation Theory developed in [1–9,13–18].

8.4 Comparison of CNT with Experiment

8.4.1 Comparison of CNT with cloud chamber experiments

A new series of condensation experiments in a diffusion cloud chamber carried out by Katz and coworkers [20–22] showed good agreement of the CNT with the measured data on the critical supersaturations of the homogeneous nucleation for the following substances: normal alkanes, n-hexane, n-heptane, n-octane, and n-nonane, for the temperature range 225–3300 K. The new Lothe–Pound theory predictions were 30%–70% lower than the measured data and the discrepancy was larger at lower temperatures.

8.4.2 Comparison of CNT results with supersonic expansion in nozzles

In a comprehensive review on the condensing flows in supersonic expanding nozzles, Wegener and Mack [23] showed that the CNT can in general agree with experiment only if the surface tension of the critical nuclei were obtained by an expression for a flat surface.

Water condensation in supersonic expansion in a highly supersaturated state was studied in [24] at different temperature gradients. The equations of motion of the condensing flow in the nozzle were solved for the given nozzle geometry, boundary, initial conditions, and centerline pressures. The heat release, mass fraction of the new phase, and the condensation rate from the condensation process were obtained from the calculated results. At temperatures from 200 to 270 K the condensate had an ice structure. One of the results was the statement that the onset of measurable condensation cannot be predicted from first principles. The main problem in predicting the condensation is the lack of the materials properties at a small number of atoms in the critical nucleus.

The CNT results were compared to experimental results of the condensing flows expanding in supersonic nozzles in [25,26]. Homogeneous nucleation in expansions of NH_3, H_2O, benzene, chloroform, Freon 11, and ethanol vapors in supersonic nozzles was studied in [25–27]. Condensation onset was detected based on the pressure measurements along the condensing flow inside the nozzle. While the results for H_2O vapor condensation in the nozzle expansion were in agreement with the CNT, without the quantum-statistical corrections introduced by Lothe and Pound [10,11], the results were inconclusive for C_2H_5OH. For the gases NH_3, benzene, chloroform, and Freon 11, the results were in good agreement with much higher nucleation rates (12 to 18 orders of magnitude) predicted by the Lothe–Pound equations. The obtained experimental data for H_2O were used for estimating the mass accommodation coefficient $0.01 < \gamma < 0.1$.

Condensation of SF6 in two continuously operating supersonic nozzle flows was studied in experiments [23]. The onset of nucleation inside the nozzle was registered by Rayleigh light scattering. The flow pressure was measured in static conditions and compared to the solution of Navier–Stokes equations for the flow. Temperature in the range of 77–113 K and flow pressures in the interval $0.09 < p < 4.6$ Torr, below the triple point of the SF6 gas, were studied. The adiabatic cooling measured in the range from 38°C to 45°C was in agreement with the CNT.

The results obtained for condensation of various gases in nozzles carried out by Wegener et al. [23] gave controversial results, although many substances show close comparison to the CNT developed by Volmer and others [1–8].

8.5   Kinetics of Condensation in Rarefied Gases*

The kinetic coefficients calculated in Chapter 7 cannot be directly compared to experiment. The comparison made in [28] was possible due to the fact that the authors solved the inverse problem and restored the kinetic coefficients from the experimental data. Therefore, in this section a systematic comparison was made of the solutions of the kinetic equations of condensation using the calculated kinetic coefficients obtained by MD with an experiment on supersonic flow (Section 8.5.2) and behind the shock wave front (Section 8.6.1).

8.5.1   Kinetics of condensation in free supersonic flow

For numerical calculation of the kinetic rate Equations 8.2, a cluster mass density is introduced q_n (1/g). In this case, the kinetic equations will be presented as follows:

$$\frac{dq_n}{dt} = J_n - J_{n+1},$$

$$(n = 3, 4, \ldots, n_{max}), \qquad (8.21)$$

$$J_{n+1} = c_n q_n q_1 \rho - e_{n+1} q_{n+1},$$

where:

ρ is the gas density

---

* The content of this chapter was published in Insepov, Z. et al., *Nucleation—Clusters—Fractals*, University of Rostock, Germany, pp. 141–153, 1991; Insepov, Z. and E. Karataev, *Molec Model*, 5(8), 48–56, 1993 (in Russian) [29]; Insepov, Z. et al., *Z Phys* D 20, 449–451, 1991.

n_{max} is a parameter of the model, the maximum size of the clusters that was varied between 100 and 200, in order to make the result independent of this parameter

Dimers were formed in three-particle collisions and the inverse process led to their destruction:

$$\frac{dq_2}{dt} = J_2 - J_3,$$

$$J_2 = k_{tr}q_1^3\rho^2 - k_{sd}q_2q_1\rho, \qquad (8.22)$$

$$q_1 = q_1^0 - \sum_{n\geq2} nq_n,$$

where q_1^0 is the density of atoms in the source.

Equations 8.22, together with the equations of gas dynamics of the jet, build a closed set of equations, the solution of which can be compared to experiment [30].

Approximation of a gas as a nonviscous and non-thermoconductive, having a one-dimensional geometry, and including the following assumptions:

1. The gas is ideal.

2. The clusters do not contribute to the pressure.

3. The velocities of the monomers and clusters are equal in the flow, the set of equations will be represented as follows:

$$u\frac{du}{dx} + \frac{(1-\mu)}{\rho}\frac{d\rho}{dx} = 0,$$

$$u\frac{du}{dx} + (1-\mu)C_p\frac{dT}{dx} = \frac{d\left(\sum_{n\geq2} dh/dn \cdot q_n\right)}{dx}, \qquad (8.23)$$

$$P = \rho RT(1-\mu),$$

where:

$\mu = \sum_{n\geq2} nq_n/q_1^0$	is the mass fraction of clusters
h	is the specific enthalpy of cluster formation
R	is the universal gas constant

The kinetic coefficients in the rate equations were calculated by MD in Chapter 7. The rate constant of a three-particle dimer formation was calculated according to the formula $k_{tr} = \alpha_{tr} 8\pi^{3/2} \sigma 5 (k_B T/m)^{1/2}$, where α_{tr} is the probability of three-particle collision that was calculated using the data and the formula $\alpha_{tr} = p_s \exp(-E_0/kT)$, where $p_s = 0.05$ and $E_0 = 0.6\varepsilon$ [31]. The specific enthalpy of cluster formation was adapted from other MD calculations [32] as follows:

$$h/\varepsilon = \begin{cases} 1.454 \ln n - 0.516, & n \leq 21, \\ 8.056\left(1 - 1.105 n^{-1/4}\right), & n > 21. \end{cases}$$

The direct nozzle problem was solved, that is, the density ρ was given along the central jet axis:

$$\rho = \rho_0 \left[\exp\left(-1.511 \bar{x}^{0.521}\right)\right]^{\frac{1}{\gamma-1}}, \quad \bar{x} \leq 3,$$

$$\rho = \frac{\rho_0}{\left(1 + \dfrac{\gamma-1}{2} M^2\right)^{\frac{1}{\gamma-1}}}, \quad \bar{x} > 3,$$

where:

$\bar{x} = x/d$

d is the diameter of the aperture

$\gamma = 5/3$,   the parameter of the adiabatic process.

The Mach number was obtained by approximation [33]:

$$M = A(\bar{x} - \bar{x}_0)^{\gamma-1} \frac{-\dfrac{1}{2}\left(\dfrac{\gamma+1}{\gamma-1}\right)}{A(\bar{x} - \bar{x}_0)^{\gamma-1}}.$$

The Mach number of the flow that is free-molecular was obtained from the expression $M_t = 133\,(p_0 d)^{0.4}$, where p_0 (atm) is the stagnation pressure and d (cm) is the aperture.

The set of Equations 8.22 and 8.23 were solved numerically by the third-order Godunov fast sweep method [34]. The following source parameters were chosen from experiment [30]: $p_0 = 5$ atm, $T_0 = 300$K, $d = 0.01$ cm.

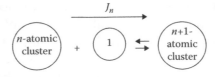

Figure 8.1   Mechanism of cluster size change by association or dissociation of a monomer.

8.5.2   Comparison with experiment

The calculated results are presented in Figure 8.1 for the mass fractions of the clusters with the sizes $n=2, 3, 4, 10$, and 20 and for the size distribution of clusters (line 1). Figure 8.1 shows that at $\bar{x} \approx 7$ the solutions approach the steady-state solution. The figure also shows good agreement between the calculated and experimental data for small clusters $n=2$–7 and a larger deviation for the sizes $n \geq 10$. The results were compared with the calculated results by Lu and Yang [28] (line 2) that used the value $\alpha_n = 1$, including the rate constant for the dimer formed in the bimolecular reaction between two monomers.

8.6   Kinetic of Condensation behind the Shock Wave Front

During shock wave propagation through a gas Ar containing small additives of pentacarbonyl of iron, a massive ejection of iron atoms occurs as a result of ionization, and the gas of metal atoms easily becomes strongly supersaturated. In [35,36], it was shown that the characteristic dissociation time calculated by MD determines the characteristic vapor condensation time, and allows comparison with experiment [37].

Dimers can be formed in the following reaction (see Chapter 2, Equation 2.11):

$$Fe + FeCO \xrightarrow{k_2} Fe_2 + CO.$$

Large clusters were able to grow from the smaller cluster by a binary reaction:

$$Fe + Fe_n \xrightarrow{k_n} Fe_{n+1}, \quad n = 2, 3, \dots \tag{8.24}$$

Neglecting evaporation from the clusters, the growth rate of the average cluster $n = \langle N_k \rangle$ can be represented as follows:

$$\frac{dn}{dt} = kn^{2/3}[\text{Fe}], \tag{8.25}$$

where $k = k_n/n^{2/3}$ is the rate constant for bimolecular association of two iron atoms $\text{Fe} + \text{Fe}$.

Equation 8.25 has the following solution:

$$n_\tau(t) = \left(\frac{1}{3}k\int_\tau^t [\text{Fe}]dy\right)^3, \tag{8.26}$$

where:

   τ is the birth time of a cluster
   t is the current time

The number of new-born clusters df_τ in a unit volume and during a time interval $d\tau$ will be

$$df_\tau = k_2[\text{FeCO}][\text{Fe}]d\tau. \tag{8.27}$$

The rate of consumption of Fe atoms can be obtained from

$$d\left(\frac{d[\text{Fe}]}{dt}\right) = -kn_\tau^{2/3}(t)[\text{Fe}]df_\tau. \tag{8.28}$$

By integrating Equation 8.28 twice and by replacing [Fe] and [FeCO] with their initial values, the time dependence of iron concentration [Fe] can be obtained as follows:

$$[\text{Fe}](t) = [\text{Fe}]_0 \exp\left(\frac{-t^4}{t_k^4}\right), \tag{8.29}$$

where t_k is the characteristic time of condensation:

$$t_k = \left(\frac{108}{k^3}k_2[\text{FeCO}]_0[\text{Fe}]_0^3\right)^{1/4}. \tag{8.30}$$

8.6.1   Comparison with experiment

Figure 8.2 shows the dependences of the characteristic time of condensation on temperature of the thermostat. The calculations were conducted with Equation 8.10, with the following parameters: $k = 3 \times 10^{-10}$ cm^3/s, $k_2 = 1.2 \times 10^{-10}\alpha_2$ cm^3/s, $[\text{Fe}]_0 = 6.73 \times 10^{15}$ cm^{-3},

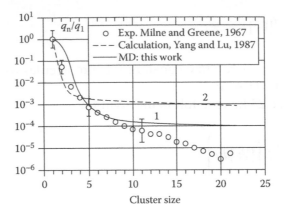

Figure 8.2   Comparison of calculated results for the size distribution of clusters (line 1) with experiment by Milne and Greene and the calculated results by Lu and Yang (line 2). (Data from T. A. Milne and F. T. Greene, *J Chem Phys* 47(10), 4095–4101, 1967; S.-N. Yang and T.-M. Lu, *Solid State Commun* 61(6), 351–354, 1987.)

$[FeCO]_0 = 3.27 \times 10^{15}$ cm^{-3}. Line 1 corresponds to the case $\alpha_2 = 1$ (complete sticking). Line 2 was obtained by using the rate constant α_2 calculated by MD in Chapter 7. Lines A, B, and C were calculated according to the quasi-chemical approximation in [38].

Figure 8.3 shows that the chemical mechanism of condensation of iron vapor behind the shock wave front significantly improves the comparison of theory with experiment, and the selection of α_2 from the MD data makes a good agreement between theory and experiment.

Since it was initially assumed that evaporation processes from clusters can be neglected, it is hoped that taking the evaporation of clusters into account would make the agreement between the theory and experiment even better.

8.7   Summary

1. The Classical Nucleation Theory (CNT) was analyzed and it was stated that one of the unresolved issues of the CNT was the difficulty of determining the kinetic coefficients. The gas dynamic formulas that were usually used to calculate the kinetic coefficients could not take into account the probabilities of association and dissociation of single atoms from clusters.

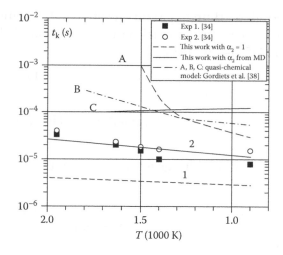

Figure 8.3   Comparison of calculated results for the characteristic time of condensation. Symbols for experimental data [39]: ■, minimal time of condensation; ○, maximal time of condensation. Line 1 corresponds to the probability of dimerization $\alpha_2 = 1$; line 2 corresponds to the calculation of α_2 by MD in this work; lines A, B, and C correspond to quasi-chemical model [37,38]. (Data from S. K. Aizatullin et al., *Chem Phys* 46(6), 851–856, 1985; Gordiets et al., *J Sov Laser Res* 7(6), 588–616, 1986.)

2. A new method of calculating the kinetic coefficients using transition state theory and molecular dynamics was developed that allows predicting the condensation processes in supersonic nozzles and behind the shock wave front.

3. The solution of the kinetic equations together with the kinetic coefficients obtained by the new method gives an excellent agreement with experiment.

References

1. J. W. Gibbs, *The Collected Works in Two Volumes*, Vol. I, *Thermodynamics*, Longmans, Green: New York, 1928; The Scientific Papers of J. W. Gibbs and Gibbs J. W.-Dover, 1, 55–353, 1961.

2. M. Volmer and A. Weber, Keimbildung in übersättigten Gebilden (Nucleus formation in supersaturated systems), *Z Phys Chem (Leipzig)* 119, 277–301, 1926.

3. M. Volmer, Bemerkung zu dem Gesetz von I. Thomson und W. Gibbs: Über den Dampfdruck von kleinen Partikeln, *Annalen der Physik*, 415, 44–46, 1935.

4. M. Volmer and H. Flood, Tröpfchenbildung in Dämpfen, *Z Phys Chem* 170, S273, 1934.

5. M. Volmer, *Kinetik der Phasenbildung*, Theodor Steinkopff: Dresden, 1939.

6. L. Farkas, Keimbildungsgeschwindigkeit in übersättigen Dämpfen, *Z Phys Chem (Leipzig)* 125, 236, 1927.

7. R. Becker and W. Döring, Kinetische Behandlung der Keimbildung in übersattigten Dämpfen, *Ann Physik* 24, 719–752, 1935.

8. J. B. Zel'dovich, *Acta Phys Chem URSS* 181, 1, 1943 (English transl.): *Soviet Phys-JETP* 12, 525, 1942.

9. J. J. Frenkel, *Kinetic Theory of Liquids*, Oxford University Press: New York, 1946.

10. J. Lothe and G. M. Pound, Concentration of clusters in nucleation and the classical phase integral, *J Chem Phys* 48, 1849–1852, 1968.

11. J. Lothe and G. M. Pound, Reconsideration of nucleation theory, *J Chem Phys* 36, 2080–2085, 1962.

12. H. Reiss, J. L. Katz, and E. R. Cohen, Translation–rotation paradox in the theory of nucleation, *J Chem Phys* 48, 5553–5560, 1968.

13. A. J. Barnard, The theory of condensation of supersaturated vapours in the absence of ions, *Proc Roy Soc A* 220(140), 132–141.

14. F. F. Abraham, *Homogeneous Nucleation Theory*, Academic Press: New York, 1974.

15. A. A. Lushnikov and A. G. Sutugin, Present state of the theory of homogeneous nucleation, *Russ Chem Rev* 45, 197, 1976.

16. J. J. Burton, Nucleation theory, in B. J. Berne (ed.), *Statistical Mechanics, Part A: Equilibrium Techniques*, pp. 195–234. Plenum: New York, 1977.

17. D. Kashchiev, Nucleation, *Basic Theory with Applications*, Butterworth/Heinemann: Oxford, 2000.

18. V. V. Slezov, *Kinetics of First-Order Transitions*, Wiley-VCH: Weinheim, 2009.

19. J. Feder, K. C. Russell, J. Lothe, and G. M. Pound, Homogeneous nucleation and growth of droplets in vapours, *Adv Phys* 15(57), 111–178, 1966.

20. J. L. Katz, Condensation of a supersaturated vapor. I. The homogeneous nucleation of the *n*-alkanes, *J Chem Phys* 52, 4733, 1970.

21. J. L. Katz, C. J. Scoppa, N. G. Kumar, and P. Mirabel, Condensation of a supersaturated vapor. II. The homogeneous nucleation of the n-alkyl benzenes, *J Chem Phys* 62, 448, 1975.

22. J. L. Katz, P. Mirabel, C. J. Scoppa II, and T. L. Virkler, Condensation of a supersaturated vapor. III. The homogeneous nucleation of CCl_4, $CHCl_3$, CCl_3F, and $C_2H_2Cl_4$, *J Chem Phys* 65, 382, 1976.

23. P. P. Wegener and L. M. Mack, Condensation in supersonic and hypersonic wind tunnels, *Adv Appl Mech* 5, 307–447, 1958.

24. P. P. Wegener and A. A. Pouring, Experiments on condensation of water vapor by homogeneous nucleation on nozzles, *Phys Fluids* 7, 362, 1964.

25. K. M. Duff, Non-equilibrium condensation of carbon dioxide in supersonic nozzles, D Sci thesis, Massachusetts Institute of Technology, 1966.

26. H. L. Jaeger, E. J. Willson, P. G. Hill, and K. C. Russell, Nucleation of supersaturated vapor in nozzles, I. H_2O and NH_3^*, *J Chem Phys* 51, 5380, 1969.

27. D. B. Dawson, E. J. Willson, P. G. Hill, and K. C. Russell, Nucleation of supersaturated vapor in nozzles, II. C_6H_6, $CHCl_3$, CCl_3F, and $C_2H_5OH^*$, *J Chem Phys* 51, 5389, 1969.

28. S.-N. Yang and T.-M. Lu, The sticking coefficient of Ar on small Ar clusters, *Solid State Commun* 61(6), 351–354, 1987.

29. Z. Insepov and E. Karataev, Calculation of the rate constants of growth and evaporation behind the shock wave front by molecular dynamics, *Molec Model*, 5(8), 48–56, 1993. (In Russian).

30. T. A. Milne and F. T. Greene, General behavior and equilibrium dimer concentration, *J Chem Phys* 47(10), 4095–4101, 1967.

31. A. P. Godfried and I. F. Silvera, Raman studies of argon dimers in a supersonic expansion. 2. Kinetics of dimer formation. *Phys Rev A* 27(6), 3019–3030, 1983.

32. A. C. Reardon and D. J. Quesnel, Growth of equilibrium clusters of Lennard-Jones molecules, *J Comput Phys* 83, 240–245, 1989.

33. Y. Ozaki, K. Murano, K. Izumi, and T. Fukuyama, Dimer concentration in supersonic molecular beams of argon and carbon dioxide, *J Phys Chem* 89(23), 5124–5132, 1985.

34. C. Hirsch, *Numerical Computation of Internal and External Flows*, vol. 2, John Wiley, 1990.

35. Z. Insepov, E. Karataev, and G. E. Norman, Kinetics of Ar cluster formation in supersonic jet, in *Nucleation—Clusters—Fractals*, F. Schweitzer and H. Ulbricht (eds), University of Rostock: Germany, pp. 141–153, 1991.

36. Z. Insepov, E. Karataev, and G. E. Norman, The kinetics of condensation behind the shock front, *Z Phys D* 20, 449–451, 1991.

37. S. K. Aizatullin, I. S. Zaslonko, V. N. Smirnov, and A. G. Sutugin, Study on condensation of iron vapor at decay in shock waves, *Chem Phys* 46(6), 851–856, 1985 (in Russian).

38. B. F. Gordiets, L. A. Shelepin, and Y. S. Shmotkin, Kinetics of isothermal processes of homogeneous condensation, *J Sov Laser Res* 7(6), 588–616, 1986.

39. T. A. Milne, A. E. Vandergrift, and F. T. Greene, Mass-spectrometric observations of argon clusters in nozzle beams. The kinetics of dimer growth, *J Chem Phys* 52(3), 1552–1560, 1970.

CHAPTER 9

Introduction to Advanced Surface Modification with Gas Cluster Ion Beams

9.1 Introduction

Physical clusters are aggregates that can consist of hundreds and thousands of gas atoms or molecules bound together via weak van der Waals forces. Most of the physical clusters are represented by rare-gas atomic clusters, such as helium, argon, xenon, or krypton. Beams of large gas clusters can be generated in supersonic expansions of a gas into vacuum through a Laval nozzle [1–5]. Electronic ionization of clusters in vacuum and their electrostatic acceleration form energetic cluster ion beams. These beams can be used to bombard a target placed in the same vacuum chamber, and can simultaneously deliver large numbers of cluster atoms at a low energy per atom, while simultaneously getting sputtering yields from the target many orders of magnitude higher than that of a monomer ion irradiation [6].

In contrast, chemical clusters are atomic groups bound together via strong chemical forces, such as covalent, hydrogen, or metallic, and are closer to large molecules by their physical and chemical properties.

Interactions of energetic gas clusters with solid surfaces demonstrate unique phenomena and promise new applications for surface modification technology [6–15]. Clusters of gaseous elements and compounds consisting of hundreds to thousands of atoms, with energies from a few eV to a few hundreds of eV per cluster atom are of particular interest for surface modification [8]. A crystal surface, with an initial average surface roughness of tens or hundreds of angstroms, becomes atomically flat, with the residual roughness of a few angstroms [6].

The study of crater formation, crater structure, and faceting properties with low cluster ion fluence are important tasks for understanding fundamental surface science and smoothing effects. An individual cluster impact would leave a crater on the surface, but it is impossible to study experimentally those craters at a high cluster ion fluence that is typical for surface smoothing processes.

The physical processes accompanying an atomic cluster ion impact on a surface are quite different from those at a single ion implantation into a target. A 1000-atom gas cluster accelerated through 10 kV yields an average individual atom energy of 10 eV as the cluster impacts the surface. This average energy is very large compared with the typical binding energy (<0.1 eV/argon atom) of the clusters, and hence the impacts are highly inelastic.

While experimental data are still sparse [6–8,10], modeling and post-atomic cluster impact analysis may help to evaluate the physical nature and mechanisms involved in gas cluster ion-beam processing. The aim of this work is to study individual Ar gas cluster impact data obtained by high-resolution transmission electron microscopy (HRTEM), atomic force microscopy (AFM), and MD simulations. High- and low-energy Ar and O_2 gas cluster impacts are observed in cross section by HRTEM. AFM is used to examine the crater surface shapes for the high-energy Ar cluster impacts for two different silicon surface orientations. The cross section and surface shape of the high-energy Ar gas cluster impact craters are compared directly with MD simulations.

In this study, polished silicon substrates with a native oxide were exposed to low fluence (10^{10} ions/cm²) argon and O_2 gas cluster ions. HRTEM cross-section images of the individual cluster impacts for Ar and O_2 at 3 and 24 kV acceleration energies were obtained. AFM images of the larger 24 kV Ar cluster impact craters were analyzed. HRTEM revealed that no dislocation formation or lattice amorphization occurs in the near-surface or subsurface regions. The GCIB smoothing process may be envisioned as a stochastic overlay of individual shallow craters. The respective shallow crater shapes of the 3 kV Ar and 3 kV O_2 cluster impacts appear to provide a fundamental basis for the resulting smoothness of material surfaces.

Unique features of cluster ion beams such as shock wave generation, lateral sputtering, and surface hardness measurement were also studied by MD simulation of a single Ar cluster impacts on copper surfaces, and the results were compared with scanning tunneling microscopy (STM) images obtained for the craters on the Au films deposited on mica [16,17].

At higher cluster ion irradiation doses of 10^{13}–10^{15} ions/cm^2, new processes such as surface smoothing and formation of surface ripples and tips will be discussed [18–25]. In this chapter a few applications of the GCIB technology will be presented, such as mitigation of high-gradient breakdown, shallow junction formation, and cluster ion–assisted thin film deposition.

9.2   Crater Formation with Hypersonic Velocity Impacts

The phenomenon of crater formation is well known in so-called hypersonic velocity (or hypervelocity) impacts of macroscopic bodies on a solid surface at velocities in the range $v_p/c > 1$, where v_p and c are the projectile and sound velocities (for target material), respectively. Crater formation at hypervelocity impacts of macroscopic projectiles on metal surfaces was studied in [26–28]. It was shown that at a hypervelocity impact, for $v_p < 10$ km/s, the crater depth fits well the empirical formula (in CGS units in the original [a misprint of the original formula was corrected in Equation 9.1]):

$$
\frac{h}{D_p} = \left(\frac{12 \times 10^{-9} \, (E/B)}{2\pi D_p^3} \right)^{1/3} ,
\tag{9.1}
$$

where:
   h is the crater depth
   D_p is the projectile diameter in cm
   E is the projectile energy in erg
   B is the standard Brinell hardness number in kg/mm^2

The shape of the macroscopic craters has been obtained to be hemispherical [26–28].

As it is seen from this formula, the crater depth h does not depend on the projectile momentum, but on the impact energy E only. For higher v_p values, the projectile momentum contributes more to the crater shape, and the exponent is slightly smaller [27]. According to this empirical formula, the projectile energy E divided by the crater volume $V_{cr} \sim h^3$ should be linearly proportional to the Brinell hardness of the target material: $E/V_{cr} \sim B$ for a hypervelocity impact with the velocity less than about 10 km/s. This correlation has been confirmed experimentally for a variety of metals including lead, aluminum, copper, bronze, brass, steel, and titanium for projectile masses ranging from 10^{-11} to 10 g (i.e., 12 orders of magnitude) for velocities up to about 15 km/s [27]. The measured crater dimensions were the depth and the radius that were obtained to be about equal.

One of the most significant effects of bombardment by heavy monomer, molecular, and cluster ion beams, with a total energy of about 10–500 keV, is the formation of atomic-scale craters, with diameters of about ~10–100 Å [18–24,29–32]. Merkle and Jäger [29] observed crater formation by TEM on Au foils due to 10–500 keV irradiation by Bi and Bi_2 ions. Thompson and Johar [30] proposed the existence of an energy threshold for crater formation with heavy monomer ion impacts above which this phenomenon can occur. The threshold energies given in [30] are 3.04 eV for Ag, 3.78 eV for Au, and 5.95 eV for Pt. These data are well correlated with the binding energies of these metals [31]. This correlation, which was observed experimentally, also shows that the dynamics of crater formation for heavy-ion impact is controlled by the total ion energy released at impact rather than by the ion momentum.

Formula (9.1) was originally obtained for large craters, with diameters of about 1 cm, created with *macroscopic* projectiles having hypersonic velocities. It is not directly applicable for *microscopic* small craters created by single heavy ions because the ions lose their energy in collisions with the target electrons. Nevertheless, there is a similarity in the physics of crater formation due to a single heavy-ion bombardment creating a track and a macroscopic body impact forming a crater.

So far it has not been shown that this formula is valid for small craters made by accelerated cluster impacts. However, in [32] this formula was used to estimate crater depths to be in the order of 20–300 Å when Cu and Ti surfaces were eroded with CO_2 and Cs clusters ($n = 100$–1000), with energies of 1–500 keV.

For the ion and surface engineering communities, a direct relationship between the physical properties of different surfaces and crater dimensions, sputtering yield, and erosion rate would have a distinctive advantage over the present state of the art: if a correlation such as Equation 9.1 could be confirmed, surface hardness could practically be obtained from other data measured routinely in cluster ion beam experiments, without performing the actual hardness measurement on another instrument.

9.3   Surface Sputtering

Surface sputtering is a process of surface erosion by energetic ion impacts. Self-sputtering is a process of surface sputtering by ions of the same kind as the target material. In this section, we will study sputtering of copper and silicon surfaces by argon cluster gas ions, with energies of up to 20 keV, and self-sputtering of a copper surface by energetic Cu^+ ions, with energies of 50 eV–50 keV.

The so-called *lateral sputtering effect*, which was predicted by molecular dynamics (MD) simulation and verified experimentally by the Kyoto group [33,34], is used to explain the surface smoothing effect that is observed for metal, semiconductor, and insulating surfaces. The MD simulation has revealed the atoms ejected from the surface have a significant lateral momentum component (parallel to the surface) which may have a major effect on surface morphology. According to the lateral sputtering effect, smoothing occurs as a cooperative result of multiple atomic-scale "bursts" caused by energetic cluster ion impacts onto surfaces. Most of the surface material involved in such bursts does not leave the surface at all but becomes highly mobile, and therefore diffuses quickly along the lateral direction, thus making the surface smoother.

Evolution of surface morphology under cluster ion irradiation was described by the modified Kuramoto–Sivashinsky equation (see Chapter 2 for more details). Comparison of the simulations with experimental data shows qualitative agreement.

Clusters are weakly bound aggregates consisting of hundreds or thousands of atoms or molecules. There is a variety of methods to prepare them. Gas clusters can easily be formed in an isentropic expansion of the gas through small nozzles. Clusters can be charged by a low-energy electron beam, producing cluster ions that may be accelerated in an electric field toward a solid target mounted in a vacuum chamber. Cluster ion machines, designed recently, irradiate solid targets by cluster beams consisting of up to several thousands of gas atoms or molecules with energies of 10–200 keV. These tools have been used for the study of new surface modification effects caused by the interaction of accelerated cluster ions with solids [35].

Understanding these complex processes is facilitated by computer simulations involving a molecular dynamics (MD) method in which equations of motion of interacting particles in a physical system are solved with appropriate boundary conditions. The number of atoms that can be treated in a system consisting of a cluster and a solid target is limited by the available computing power. Thus the solid is represented by a finite number of atoms and the rest of the system by a continuum with the proper boundary between the two regions, which allows the flow of energy deposited by the impact to the bulk of the solid.

The aim of this work was to study sputtering caused by energetic cluster ion bombardment of a metal surface by MD. The projectiles are Ar clusters consisting of 200–1000 atoms, with total cluster energies of 6–20 keV, corresponding to ~20 eV per cluster atom. That energy was chosen in order to make a comparison of the MD results with the experimental sputtering yields measured in our laboratory for Ar3000 cluster ions bombarding Cu and Ag targets, with energies in the range of 10–25 keV [36,37].

We have also compared our MD results with experimental point at 30 keV for Ar300 cluster ion bombardment of a gold surface [6]. We have also studied the evolution of the morphology of a surface irradiated by energetic cluster beams by using the Kuramoto–Sivashinsky (KS) equation.

9.4   Shock Wave Generated by Energetic Cluster Impacts (2D Case)

Shock waves occur at macroscopic body impacts on planets or explosions [38,39], laser ablation of solid surfaces [40,41], high-energy ion impact on a ICF targets [42,43], and large organic molecule impacts [7,8], and are of great scientific and technological interest [44].

Theoretical descriptions of shock damage are based on a set of hydrodynamic equations related to mass, momentum, and energy conservation laws, as well as an additional equation of state of a material, and a kinetic equation of disordering and defect accumulation [38,39,45]. An important result of this theory is the self-similarity of shock waves. It is a matter of fact that shock waves generated at different impacts behave similarly at later time intervals [44,45]. The parameters of self-similarity can be found from numerical solutions of the hydrodynamic equations or from experiment. In [46] the authors studied the microscopic shock waves generated in heavy-ion bombardment of high-Z materials assuming that the thermal equilibrium could be achieved during the impact. The experiment deals with so-called high-velocity impacts of macroscopic bodies at velocities in the range v_p/c ~1–10, where v_p and c are the projectile and sound velocities, respectively. There are several empirical relations available [38] that can predict crater depth:

$$\frac{h}{d_p} = K \left(\frac{\rho_p}{\rho_t} \right)^{2/3} \left(\frac{v_p}{c} \right)^{\beta},$$

(9.2)

where:
- h is the crater depth
- d_p is the diameter of the projectile
- K is a numeric constant
- ρ_p and ρ_t are the projectile and the target densities, respectively

The values of the exponent β are between 2/3 and 1/3, depending on whether the crater volume is proportional to the energy or momentum released in the impact, respectively. A value $\beta \cong 0.58$ was found experimentally [38].

Laser ablation shocks were found to possess the same features as shock waves resulting from high-velocity impacts [41]. It has been shown that the position of the shock front as a function of time obeys the self-similarity form: $R = At^{\alpha}$, where R is the distance traveled by the shock front in time t. This experiment gave a value $\alpha = 0.61$ for the exponent which is close to the prediction of hydrodynamic theory for a planar shock front.

Accelerated cluster-ion impacts on solid targets created a new field of application of shock wave theory. Such impacts generate, in fact, a new type of atomic or microscopic shocks in a target. It is not obvious that these microscopic shock waves, which are generated at a very short time scale (less than ~1–5 ps) and on a very small space scale (in the order of ~10 Å), have the same properties as conventional macroscopic shocks.

The aim of this chapter is to study shock wave generation at cluster–surface impacts by the two-dimensional (2D) molecular dynamics (MD) method. This method has been used widely before for the simulation of the deposition of slow clusters on crystal surfaces [11,24,47], generation of shock waves inside the cluster hitting a rigid surface [48], and sputtering of target material due to cluster bombardment [49]. The structure of planar shocks in solid Ar was studied by this method in an earlier paper [50]. The shock waves generated by cluster impacts on solid surfaces have not been studied by the MD method before.

9.5   Surface Smoothing

One of the new remarkable effects of surface sputtering with gas cluster ion beams is surface smoothing. To understand its mechanism we have developed a computational model, which utilizes molecular dynamics (MD) to simulate rapid atomic collisions in the central impact zone and a finite-difference method to account for processes occurring on a longer scale over a wider target area. This model is discussed in more detail in Chapter 2. A case of gas cluster from a few tens to a few thousands Ar atoms impacting a Cu target with energy up to 20 keV has been considered.

The ultimate goal of the advanced surface technology is to develop a new surface polishing method capable of achieving atomic-scale surface roughness (<1 Å). Surface roughness can be characterized by absolute deviation of the surface from the average (R_a) and root mean square (*rms*) roughness, which are simple measures of the height profile roughness and are defined as follows:

$$R_a = \frac{1}{N} \sum_{i=1}^{N} \left| h(x_i) - \overline{h} \right|,$$

$$rms = \sqrt{\frac{1}{N} \sum_{i=1}^{N} \left(h(x_i) - \overline{h} \right)^2}, \qquad (9.3)$$

$$\overline{h} = \frac{1}{N} \sum_{i=1}^{N} h(x_i),$$

where:
  N is the total number of data points
  $h(x_i)$ is the height at point x_i
  $\overline{h}$ is the average of the heights

The R_a and *rms* values depend on the measurement area: the larger the measurement area, the larger the R_a and *rms*.

Therefore, to compare the data obtained by different methods, the measurement areas should be as close to each other as possible. The following areas will be used in this chapter for comparison between different methods: 1×1, 5×5, 10×10, and 20×20 μm^2.

Figure 9.1 compares the average surface roughness values obtained from the data given in Table 9.1. Table 9.1 shows the experimental data available in the literature for advanced and recently developed surface polishing techniques, which include chemical mechanical polishing (CMP), tribochemical polishing (TCP), superpolishing (SP) of astrophysical mirrors, and GCIB surface smoothing.

Chemical–mechanical, mechano-chemical, or hydrodynamic polishing methods (CMP) [51–64] are widely used in optical and semiconductor manufacturing. CMP uses mechanical and hydrodynamic pressure delivered to a workpiece (a sample surface) by polymer pads and a fluid containing abrasive fine particles. A chemically active solution (KOH, acids) containing hard oxide particles as abrasives is used for processing. CMP mostly operates through oxidation of the material and removal of the product by abrasion. The workpiece (a wafer) is held by a rotating carrier

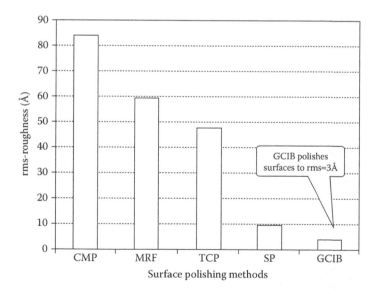

Figure 9.1   Surface roughness obtained with various surface polishing techniques. SP is a technique that combines metal electroplating with magnetron sputtering deposition, that is, it is not a true smoothening process but rather a multistage process that includes various methods. More details for the roughness data are given in Table 9.1.

Table 9.1 Comparison of Various Surface Polishing Techniques

Materials	Chemical Mechanical Polishing (CMP)			Tribochemical Polishing (TCP)			Superpolishing for X-Ray Mirrors (SP)			Gas Cluster Ion Beam Polishing (GCIBP)		
	PR (μm/h)	Ra (Å)	Reference Comments	PR (μm/h)	Ra (Å)	Reference Comments	PR (μm/h)	Ra (Å)	Reference Comments	PR (μm/h)	Ra (Å)	Reference Comments
AlGaAs	5	2	[52] Single crystal									
Si₃N₄	0.06–7.2	9; 166	[57]; [58,63]	1–3	5	[65]					1.7	1×1 μm²
SiNₓ								8–11	[78]			
SiC	0.02–0.3; 0.2; 5; 7.11 [56] [53] 30×30 μm² Single crystal	<50	[59]	2–4	10	[66,67]						
Diamond	0.5	2	[62]	<1	<10	[68]						
Tungsten	20; 140; 9–11	—; —; 20	[64]; [61]; [61]	—	90	[69]						
CaF₂											1.7	5×5 μm²
Glass											2.5	20×20 μm²
Ta											4	10×10 μm²
Si	≤1	≤1	[55] 1×1 μm								<2	10×10 μm²

pressed against a rotating polishing pad. The polishing slurry containing active fluid and abrasives is distributed across the pad and reacts with the workpiece surface. Single-crystal surfaces were much flatter and smoother originally. For example, commercially available Al_xGa_{1-x} As single crystals have (111) surface roughness of about 20 Å, which could be reduced by CMP up to 2 Å [52]. A SiC single crystal was treated by CMP after damaging the surface to the depth of about 700 Å by mechanical polishing [53]. The final roughness was 7.11 Å for the scan area of 30×30 μm. Planarization of silicon oxide surfaces by CMP is discussed in [54], and a further development of CMP, called wet cleanings post-CMP, is given in [55] for silicon wafers. However, a limitation of CMP is the contamination problem. The latter paper states that the CMP method leads to a significant surface contamination of silicon by metals such as Fe, Zn, Ca, and Al, and an additional cleaning process is necessary. The results obtained for a single-crystalline surface are not applicable for polishing polycrystalline materials, because the grain boundaries react much faster than the grain body, and that reaction would make surfaces even rougher after CMP. Since all normal metals are polycrystalline, fast etching of the grain boundaries is the main reason why CMP is rarely applied to metal surface polishing. Tribochemical polishing (TCP) is a novel technique in which the workpiece is rubbed against a polishing pad made of hard material with an additional active fluid that does not contain abrasive particles [65–68]. This method is based on tribochemistry and consists in the dissolution of material stimulated by friction at contacting asperities. The pad materials are cast iron, stainless steel, and ceramics. This method has been successfully applied to smooth Si_3N_4, SiC, and tungsten. As it uses hard polishing pads and a polishing solution free of abrasive particles, no mechanical processes such as plowing, fracture, or plastic deformation are involved. The technique produces defect- and stress-free, ultrasmooth surfaces. TCP has demonstrated removal rates up to 3 μm/h and residual roughness of 10 Å. Tungsten is softer than many ceramic materials. A chromium oxide solution gives the best combination of polishing rate and surface roughness of about 90 Å [69]. A problem with the TCP method is that it uses chemical solutions that certainly would contaminate the treated surface. This method thus seems to be inapplicable to certain important industrial and technology needs such as mirror fabrication, high-voltage electrode smoothening, and silicon microelectronics. For polishing silicon nitride, the best results were obtained with 3 wt.% chromic acid [69].

The magneto-rheological polishing (MRF) method is a new polishing process that was developed in the mid-1990s and transferred to industry in 1997. MRF delivers normal and shearing forces to the surface of the workpiece and thus has many similarities to the TCP and CMP methods. Although many theories of small-scale fracture, plastic scratching, and hydrated layers have been developed, beginning as early as 1927, mechanisms governing MRF processes are not yet fully understood [63,70–77]. This method is based on the magneto-hydrodynamic behavior of magnetic and nonmagnetic particles in an aqueous or nonaqueous fluid and a magnetic field. For example, 6 vol.% nonmagnetic cerium oxide powder as an abrasive, with a median size of about 3 μm, in an aqueous suspension and 36 vol.% of carbonyl iron (CI) powder, with a size of about 4 μm, as a magnetic component, in de-ionized water and a fluid stabilizer has been found appropriate for almost all soft and hard optical glasses and low-expansion glass-ceramics. Another suspension containing nano-diamond powder is better suited to hard materials like calcium fluoride, infrared glasses, hard single crystals (silicon and sapphire), and very hard polycrystalline ceramics (silicon carbide). Finishing rates are 2–9 μm/min, depending on the hardness of the workpiece material. It is difficult to estimate the residual roughness of MRF because this method primarily uses the highest removal rate, and therefore it has deliberately not been focused on attaining the lowest surface roughness. As the method uses suspensions of powder materials with minimum size of at least 40 Å (for diamond nano-powder), this size would likely limit the MRF method. Peak-to-valley (p-v) roughness values less than 50 Å have been set as a future goal for MRF [76,77].

In astrophysics, crowded regions of the sky cannot be studied without using multilayer mirrors for focusing x-rays in the 40–100 keV energy region. Such x-ray telescopes use the focusing abilities of multilayers at grazing angles for increasing sensitivity and resolution. The synchrotron source characteristics and the grazing x-ray incidence geometry set very high requirements on the mirror surface. Perfectly smoothed surfaces are needed to achieve high reflectivity at grazing angles of incidence of x-rays in the energy range between 1 and 50 KeV. Local deviations from ideal flat surfaces should be kept within 1 μkrad rms range over the whole mirror surface area of ≈1 m and micro roughness should be within a few Å. The multilayer mirrors were fabricated by an unbalanced magnetron sputtering technique involving a 500-nm amorphous thin SiN_x film on a Ni surface (rms ~10–20 Å in the 100–1 mm wavelength range), with an additional release layer of Cu (50 nm), reflecting layer of W/Si, and electro-formed Ni layer on the top. The roughness of SiN_x was approximately of ~8–11 Å rms for the same range, measured by an optical profiler [78].

9.6 Thin Film Surface Hardness

Hardness is commonly described as the resistance of a material to the penetration of an indenter. Teter and Hemley [79,80] refer to the known difficulties of hardness testing and interpretation by a quote from a book published in 1934 [81]: "Hardness, like the storminess of seas, is easily appreciated but not readily measured." Until the relatively recent introduction of the nanohardness method [82] in which the hardness is obtained continuously in a load–unload cycle, hardness evaluation was performed by methods in which an indentor penetrates the surface at a known load. Hardness values are obtained by measuring the surface deformation after removing the indenter.

The hardness is determined either by measuring the shape or the depth of the residual indentation. Hardness values that can be directly correlated with physical properties, such as tensile strength, are obtained in techniques such as Brinell, Vickers, or Knoop, in which the surface geometry of the residual indentation is measured by optical methods. Figure 9.2 shows a schematic representation of the Brinell hardness test. The basic formula for hardness is P/A, where P is the load and A is the surface area of the indentation. For a Brinell hardness number (BHN), the formula is as follows:

$$B = \frac{2P}{\pi D\left(D - \sqrt{(D^2 - d^2)}\right)}, \tag{9.4}$$

where:

D is the indenter's diameter

d is the diameter of the residual indentation as shown in Figure 9.2 [83]

Hardness values thus obtained can be directly expressed in Pascals, as the load is known and the surface area of the indentation

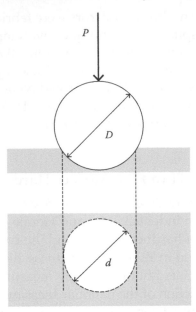

Figure 9.2   The Brinell test's force diagram: P, applied force; D, the diameter of intender; d, the diameter of indentation.

is obtained from its diameter. Figure 9.2 shows schematics of the Brinell test.

A significant problem encountered in hardness measurements is the measurement of treated surfaces and coatings. The indentation trace in this case is significantly influenced by an elastic contribution from the substrate material. This effect is reduced with increasing coating thickness, but it can vary significantly with substrate. Vickers hardness values of TiN films deposited simultaneously on various substrates [84] were found to vary at low loads between 25,000 MPa on hard steel and 4,000 MPa on Al. The fact that the same type of thin film can present different values on different substrates means that in most cases hardness values can only be used for comparison and that they are not absolute. This also leads to some doubts regarding the accuracy of much of the data currently at hand in the literature.

A significant improvement in the field of hardness measurements was the development of the nanohardness (or ultramicrohardness) testing method in which the load versus penetration depth curve is recorded during the penetration of a conical indentor into the surface at very low load [82], and data from the near-surface region is collected as a function of depth and thus it was believed to be less influenced by the substrate. Nevertheless, problems in the interpretation of results obtained from hard deposited coatings are still common, and the technique is not free from substrate effects.

It seems that at present there is no method that is reliable for absolute hardness values, particularly for thin film coatings, confirming the above-mentioned view expressed in [81]. A solution to this problem was proposed in [85] where a new concept of absolute hardness was introduced, which has been made feasible with recent developments in surface microscopy [surface tunneling microscopy (STM) and atomic force microscopy (AFM)] and in cluster ion beam technology.

9.7 Ripple and Quantum Dot Formation by Cluster Ion Impacts

Metal surfaces have a high surface mobility of adatoms and this feature has a strong influence on surface modification processes caused by ion bombardment [20,86–89]. Ripple formation on the surface of ion irradiated Si was experimentally observed [90], and on amorphous SiO_2 and crystalline Ge (001) surfaces for low (below 1 keV) energy ions heavy (Xe) and light (H, He) ions [91,92]. Ripple formation was further observed on an amorphous $Fe_{40}Ni_{40}B_{20}$ film surface [93], and ion-induced stresses and subsequent viscous flows of SiO_2 surfaces were measured directly in [94]. A two-dimensional (2D) macroscopic ripple formation theory was proposed in [95]. According to this, a uniform ion irradiation of a rough surface increases surface roughness because trough atoms gain a higher energy than the atoms of the crests. This idea was used to develop a scaling theory of ripple formation in [96]. Monte Carlo computer models for ripple formation in ion beam sputtering were recently developed in [97].

The large gas cluster ion irradiation of a surface is subjected to different mechanisms compared to single ion impact. The cluster itself does not penetrate into the substrate on impact. The model of surface instability [96] may not be valid for surface processes activated by cluster ion irradiation.

Experimental studies of surface modification with cluster ion beams clearly show that a very high smoothing rate of rough substrate surfaces with normal cluster beams could be achieved [20,98]. Surfaces may roughen under oblique cluster impacts resulting in a ripple structure. An AFM study of a Cu (100) surface irradiated with large 20 keV Ar_{3000}^+ cluster ions with doses of 5×10^{15} ions/cm^2 at 60° off-normal incidence angle shows the formation of ripples with an average wavelength of 300 Å [98,99].

The result of simulation of a large Ar–cluster impact on a Cu surface is presented in this chapter. The results of a 2D MD simulation of rough surface erosion with cluster impacts were compared with the predictions of [95]. A numerical model of ripple formation on a surface under oblique cluster ion beam irradiation is presented.

9.8   Surface Breakdown in High-Gradient Accelerators

Higher voltage gradients are required for developing future TeV linear accelerators and muon-muon colliders. However, the development of these accelerators and colliders is seriously hampered by the breakdown of electric vacuum between the electrodes in the accelerator structures, power sources, and waveguides. Electrical breakdowns in vacuum, for both RF and DC systems, have been studied for many years, and many mechanisms have been proposed to explain this phenomenon [100–104]. Breakdowns seem to have a sharp threshold in electric field, but somewhat weak dependence on frequency, temperature of structure, or vacuum conditions [105,106].

The most serious effect on the breakdown field was produced surface imperfection: contamination, foreign atoms, surface roughness, cracks, and tips could significantly reduce the critical field and thus cause the breakdown more easily. Therefore, surface smoothing has a significant mitigation effect for the mitigation of vacuum high-voltage breakdowns in vacuum systems.

Recent experiments have shown that breakdowns occur when the tensile stress exerted on the surface curvature by a local electric field becomes comparable to the tensile strength of the metal [106]. While single atoms may not interact strongly with the electric fields, the emission of clusters and fragments, in the presence of field emitted electron beams, may produce very high local power densities (10^{13}–10^{14} W/cm^3), which could rapidly ionize and damage the structure [107]. This chapter describes the effect of surface smoothing on the mechanisms of surface breakdowns. A new mechanism was proposed where a nanoscale copper tip on an interior surface of an rf cavity could be disconnected from the surface more easily than a single atom and thus initiate surface breakdown [85]. The MD simulations in this chapter describe the surface conditions and estimate the temperature and electric field dependence of this mechanism.

9.9   Engineering of Nozzle Design by DSMC

Gas clusters are assemblies of a few tens to thousands of atoms loosely bound by van der Waals forces. The gas clusters are formed in a supersonic gas expansion through a nozzle by homogeneous condensation. The gas dynamic nozzle is an important part of a machine for producing large gaseous clusters in supersonic gas expansion into vacuum. The current state of the research and development of a new gas cluster ion beam process technology has created a considerable need for optimized supersonic nozzles that would generate a high current beam with predictable cluster size distribution.

After selection, ionization, and acceleration to a high voltage, these cluster ions could be used for various novel surface processes like surface smoothing and modification, ion implantation, or high-quality film formation at low surface temperature.

The design and fabrication of a supersonic nozzle with a predictable cluster distribution function is a very important fundamental and practical problem. The theoretical study of condensation phenomena in a supersonically expanding gas is known to be the most difficult problem of classical thermodynamics.

Cluster formation in a supersonic nozzle flow is a challenging problem for both theory [108–114] and experiment [115–122]. It is well known that the gas dynamics theory based on Navier–Stokes equations cannot in principle treat this problem due to a highly nonequilibrium nature of the cluster formation phenomenon.

The direct simulation Monte Carlo (DSMC) method [123–127] is a probabilistic method, and the number of simulating "particles" in a DSMC model could be many orders of magnitude less than the real molecules' number. This method has significantly extended the space and time simulation scales, and it is a very promising future technique. Another advantage of DSMC as compared with the continuum fluid mechanics Navier–Stocks equations is that it could treat supersonic expansion of gases with inherent nonequilibrium processes, for example, with chemical reaction [18], and it can in principle take into account the formation of small clusters [127].

References

1. O. Abraham, J. H. Binn, B. G. DeBoer, and G. D. Stein, Gasdynamics of very small Laval nozzles, *Phys Fluids* 24, 1017–1031, 1981.

2. B. M. Smirnov and A. Ju Strizhev, Kinetics of clustering processes in expanding vapor, *Physica Scripta* 49, 615–618, 1994.

3. B. M. Smirnov, Model of clusterization in free jet expansion, *Physica Scripta* 50, 364–367, 1994.

4. B. K. Rao and B. M. Smirnov, Kinetics of clusterization in expanding vapor, *Physica Scripta* 56, 439–445, 1997.

5. B. M. Smirnov, Generation of cluster beams, *Physics—Uspekhi* 46(6), 589–628, 2003.

6. I. Yamada, J. Matsuo, Z. Insepov, D. Takeuchi, M. Akizuki, and N. Toyoda, Surface processing by gas cluster ion beams at the atomic (molecular) level, *J Vac Sci Technol A* 14(3), 781–785, 1996.

7. J. -H. Song, S. N. Kwon, D. -K. Choi, and W. -K. Choi, Assessment of an ionized CO_2 gas cluster accelerator, *Nucl Instrum Methods Phys Res B* 179, 568, 2001.

8. I. Yamada, J. Matsuo, N. Toyoda, and A. Kirkpatrick, Materials processing by gas cluster ion beams, *Mater Sci Eng R* 34, 231, 2001.

9. P. von Blanckenhagen, A. Gruber, and J. Gspann, Atomic force microscopy of high velocity cluster impact induced nanostructures, *Nucl Instrum Methods Phys Res B* 122, 322, 1997.

10. A. Gruber and J. Gspann, Nanoparticle impact micromachining, *J Vac Sci Technol B* 15, 2362, 1997.

11. H. Hsieh, R. S. Averback, H. Sellers, and C. P. Flynn, Molecular-dynamics simulations of collisions between energetic clusters of atoms and metal substrates, *Phys Rev B* 45, 4417, 1992.

12. Z. Insepov and I. Yamada, Computer simulation of crystal surface smoothing by accelerated cluster ion impacts, *Mater Sci Eng A* 217, 89, 1996.

13. L. P. Allen, Z. Insepov, D. B. Fenner, C. Santeufemio, W. Brooks, K. S. Jones, and I. Yamada, Craters on silicon surfaces created by gas cluster ion impacts, *J Appl Phys* 92, 3671–3678, 2002.

14. Z. Insepov, L. P. Allen, C. Santeufemio, K. S. Jones, and I. Yamada, Computer modeling and electron microscopy of silicon surfaces irradiated by cluster ion impacts, *Nucl Instrum Methods Phys Res B* 202, 261–268, 2003.

15. Z. Insepov, L. P. Allen, C. Santeufemio, K. S. Jones, and I. Yamada, Crater formation and sputtering by cluster impacts, *Nucl Instrum Methods Phys Res B* 206, 846–850, 2003.

16. D. Takeuchi, K. Fukushima, J. Matsuo, and I. Yamada, Study of Ar cluster ion bombardment of a sapphire surface, *Nucl Instrum Methods Phys Res B* 121, 493, 1997.

17. T. Seki, T. Kaneko, D. Takeuchi, T. Aoki, J. Matsuo, Z. Insepov, and I. Yamada, STM observation of HOPG surfaces irradiated with Ar cluster ions, *Nucl Instrum Methods Phys Res B* 121, 498, 1997.

18. I. Yamada, W. L. Brown, J. A. Northby, and M. Sosnowski, Surface modification with gas cluster ion beams, *Nucl Instrum Methods Phys Res B* 99, 223, 1993.

19. Z. Insepov, M. Sosnowski, and I. Yamada, Molecular dynamics simulation of metal surface sputtering by energetic rare-gas cluster impact, in I. Yamada et al. (ed.), *Laser and Ion Beam Modification of Materials*, p. 111. Elsevier: Amsterdam, 1994.

20. I. Yamada, J. Matsuo, Z. Insepov, and M. Akizuki, Surface modifications by gas cluster ion beams, *Nucl Instrum Methods Phys Res B* 106, 165, 1995.

21. T. Aoki, J. Matsuo, Z. Insepov, and I. Yamada, Molecular dynamics simulation of damage formation by cluster ion impact, *Nucl Instrum Methods Phys Res B* 121, 49, 1997.

22. I. Yamada and J. Matsuo, Gas cluster ion beam processing for ULSI fabrication, in K. N. Tu, et al. (eds), *Advanced Metallization for Future ULSI*, MRS Symposia Proceedings No. 427, p. 263. Materials Research Society: Pittsburgh, 1996.

23. R. S. Averback and M. Ghali, A rutherford backscattering-channelling study of yttrium-implanted stainless steel before and after oxidation, *Nucl Instrum Methods Phys Res B* 90, 191, 1994.

24. H. Haberland, Z. Insepov, and M. Moseler, Molecular-dynamics simulation of thin-film growth by energetic cluster impact, *Phys Rev B* 51, 11061, 1995.

25. Z. Insepov, R. Manory, J. Matsuo, and I. Yamada, Proposal for a hardness measurement technique without indentor by gas-cluster-beam bombardment, *Phys Rev* 61, 8744–8752, 2000.

26. J. K. Dines and J. M. Walsh, in R. Kinslow (ed.), *High-Velocity Impact Phenomena*, Chapter 3, p. 45. Academic: New York, 1970.

27. J. W. Gehring, in R. Kinslow (ed.), *High-Velocity Impact Phenomena*, Chapter 9, p. 463. Academic: New York, 1970.

28. D. Maxwell, Simple Z-model of cratering, ejection and over-turned flap in D. J. Roddy et al. (eds), *Impact and Explosion Cratering*, p. 1003. Pergamon: New York, 1977.

29. K. L. Merkle and W. Jäger, Direct observation of spike effects in heavy-ion sputtering, *Philos Mag* 44, 741, 1981.

30. D. A. Thompson and S. S. Johar, Nonlinear sputtering effects in thin metal films, *Appl Phys Lett* 34, 342, 1979.

31. R. Behrisch (ed.), *Sputtering by Particle Bombardment I, Physical Sputtering of Single-Element Solids*, Springer: Berlin, 1981.

32. J. Gspann, Series C: Mathematical and physical sciences, in P. Jena et al. (eds), *Physics and Chemistry of Finite Systems: From Clusters to Crystals*, vol. 374 of NATO advanced study institute, p. 1115. Kluwer: Dordrecht, 1992.

33. Z. Insepov, M. Sosnowski, and I. Yamada, Molecular dynamics simulation of metal surface sputtering by energetic rare-gas cluster impact, *Trans Mater Res Soc Jpn* 17, 111, 1994.

34. I. Yamada and J. Matsuo, Lateral sputtering by gas cluster ion beams and its applications to atomic level surface modification, *Mater Res Soc Symp Proc* 396, 149, 1996.

35. I. Yamada, A short review of ionized cluster beam technology, *Nucl Instrum Methods Phys Res B* 99, 240, 1995.

36. J. Matsuo, N. Toyoda, M. Akizuki, and I. Yamada, Sputtering of elemental metals by Ar cluster ions, *Nucl Instrum Methods B* 121, 459, 1997.

37. N. Toyoda. J. Matsuo, and I. Yamada, The sputtering effects for cluster ion beams, *Proceedings of the 14th International Conference on Application of Accelerators in Research Industry*, Denton, TX, 1996.

38. R. Kinslow (ed.), *High-Velocity Impact Phenomena*, Academic: New York, 1970.

39. D. J. Roddy, R. O. Pepin, and R. B. Merill (eds), *Impact and Explosion Cratering*, Pergamon: New York, 1977.

40. H. Szichman and S. Eliezer, Scaling laws for pressure, temperature, and ionization with two-temperature equation-of-state effects in laser-produced plasmas, *Laser Partic Beam* 10, 23–40, 1992.

41. L. J. Dhareshwar, P. A. Paik, T. C. Kaushik, and H. C. Pant, Study of laser-driven shock wave propagation in Plexiglas targets, *Laser Partic Beam* 10, 201, 1992.

42. R. Beuler and L. Friedman, Larger cluster ion impact phenomena, *Chem Rev* 86, 521, 1986.

43. C. Deutch, Ion cluster interaction with cold targets for ICF: Fragmentation and stopping, *Laser Partic Beam* 10, 217, 1992.

44. Ya. B. Zel'dovich and Yu. P. Raiser, *Physics of Shock Waves and High-Temperature Hydrodynamic Phenomena*, Academic: New York, 1967.

45. S. I. Anisimov et al., Physics of the damage from high-velocity impact, *JETP Lett* 39, 8 1984.

46. Y. Kitazoe, N. Hiraoka, and Y. Yamamura, Hydrodynamical analysis of non-linear sputtering yields, *Surf Sci* 111, 381, 1981.

47. M. H. Shapiro and T. A. Tombrello, Simulations of core excitation in energetic cluster impacts on metallic surfaces, *Nucl Instrum Methods B* 66, 317, 1992.

48. U. Even, I. Schek, and J. Jortner, High-energy cluster—surface collisions, *Chem Phys Lett* 202, 303, 1993.

49. Z. Insepov, M. Sosnowsi, and I. Yamada, Molecular dynamics simulation of metal surface sputtering by energetic rare-gas cluster impact, *Proceedings of the IUMRS International Conference on Advanced Materials*, Tokyo, Japan, 1993.

50. V. Yu. Klimenko and A. N. Dremin, Structure of a shock-wave front in a solid, *Sov Phys Dokl* 25, 288, 1980.

51. J. M. Steigerwald, S. P. Murarka, and R. J. Gutmann, *Chemical Mechanical Planarization of Microelectronic Materials*, p. 336. Wiley: New York, 1997.

52. Y. Sawafuji, J. Nishizawa, AlxGa1-xAs (111)A substrate with atomically flat polished surface, *J Electrochem Soc* 146, 4253–4255, 1999.

53. E. K. Sanchez et al., Assessment of polishing-related surface damage in silicon carbide, *J Electrochem Soc* 149, G131–G136, 2002.

54. R. Schmolke et al., On the impact of nanotopography of silicon wafers on post-chemical mechanical polished oxide layers *J Electrochem Soc* 149, G257, 2002.

55. A. Abbadie, F. Crescente, and M. N. Semeria, Advanced wet cleanings post-CMP: Application to reclaim wafers, *J Electrochem Soc* 151, G57–G66, 2004.

56. L. Zhou, V. Audurier, P. Pirouz, and J. A. Powell, Chemomechanical polishing of silicon carbide, *J Electrochem Soc* 144, L161–L163, 1997.

57. Y. Z. Hu, G. -R. Yang, T. P. Chow, and R. J. Gutmann, Chemical-mechanical polishing of PECVD silicon nitride, *Thin Solid Films* 290/291, 453–457, 1996.

58. M. Raghunandan, N. Umehara, and R. Komanduri, *Proceedings of the ASME/STLE Tribology Conference*, 30, pp. 81–97, 1994.

59. M. Kikuchi, Y. Takahashi, T. Suga, S. Suzuki, and Y. Bando, Mechanochemical polishing of silicon carbide single crystal with chromium(III) oxide abrasive, *J Am Ceram Soc* 75, 189–194, 1992.

60. D. L. Hetherington et al., *In Proceedings of the MRS Symposium*, 10, pp. 41–48, 1994.

61. I. Kim, K. Murella, J. Schlueter, E. Nikkel, J. Traut, and G. Castleman, Optimized process developed for Tungsten CMP, *Semicond Intl* 119–123, 1996.

62. J. Kuennle and O. Weis, Mechanochemical superpolishing of diamond using $NaNO_3$ or KNO_3 as oxidizing agents, *Surf Sci* 340, 16–22, 1995.

63. S. R. Bhagavatula and R. Komanduri, On chemo-mechanical polishing of silicon nitride with chromium oxide abrasive, *Phil Mag A* 74, 1003–1017, 1996.

64. O. Khaselev and J. Yahalom, Constant voltage anodizing of Mg–Al alloys in $KOH–Al(OH)_3$ solutions, *J Electrochem Soc* 145, 190–196, 1998.

65. S. R. Hah, C. B. Burk, and T. E. Fischer, Surface quality of tribochemically polished silicon nitride, *J Electrochem Soc* 146(4), 1505–1509, 1999.

66. Z. Zhu, V. A. Muratov, and T. E. Fischer, Tribochemical polishing of silicon carbide in oxidant solution, *Wear* 225/229, 848–856, 1999.

67. H. Tomizawa and T. E. Fischer, Friction and wear of silicon nitride and silicon carbide in water: Hydrodynamic lubrication at low sliding speed obtained by tribochemical wear, *ASLE Trans* 30, 41–46, 1987.

68. J. Haisma, F. J. H. M. Van der Kruis, J. M. Oomen, and F. M. J. G. Fey, Damage-free tribochemical polishing of diamond at room temperature: A finishing technology, *Prec Eng* 14, 20–27, 1992.

69. V. A. Muratov and T. E. Fisher, Tribochemical polishing. *Annu Rev Mater Sci* 30, 27–51, 2000.

70. N. Umehara and K. Kato, Fundamental properties of magnetic fluid grinding with a floating polisher, *J Magnet Mag Mater* 20/24, 428–431, 1992.

71. T. H. C. Childs and H. J. Yoon, Magnetic fluid grinding cell design, *CIRP Ann* 41, 343–346, 1992.

72. I. V. Prokhorov, W. I. Kordonski, L. K. Gleb, G. R. Gorodkin, M. L. Levin, New high precision magnetorheological instrument-based method of polishing optics, in *Optical Fabrication and Testing*, OSA Technical DigestSeries, vol. 24, p. 134, Optical Society of America, Washington, DC, 1992.

73. Y. Fuqian, D. Golini, D. H. Raguin, and S. D. Jacobs, Planarization of gratings using magnetorheological finishing, *Proc MRS Symp* 477, 131–136, 1997.

74. W. I. Kordonski and S. D. Jacobs, Magnetorheological finishing, *Intl J Mod Phys B* 10, 2837, 1996.

75. W. I. Kordonski and S. D. Jacobs, Elements and devices based on magneto-rheological effect, *J Intell Mater Sys Struct* 7, 131–137, 1996.

76. A.B. Shorey, Understanding the mechanism of glass removal in magnetorheological rinishing (MRF), *LLE Rev* 83, 157–172, 2000.

77. J. E. DeGroote, H. J. Romanofsky, I. A. Kozhinova, J. M. Schoen, and S. D. Jacobs, Polishing PMMA and other optical polymers with magnetorheological finishing, *LLE Rev* 96, 239–249, 2003.

78. M. P. Ulmer, R. Altkorn, M. E. Graham, A. Madan, and Y. S. Chu, Production and performance of multilayer-coated conical x-ray mirrors, *Appl Optics* 42, 6945–6952, 2003.

79. D. M. Teter and R. J. Hemley, Low-compressibility carbon nitrides, *Science* 271, 53, 1996.

80. D. M. Teter, Computational alchemy: The search for new superhard materials, *MRS Bull* 23(1), 22, 1998.

81. H. O. O'Neill, *The Hardness of Metals and its Measurement*, Chapman and Hall: London, 1934.

82. D. Newey, M. A. Wilkins, and H. M. Pollock, An ultra-low-load penetration hardness tester, *J Phys E* 15, 119, 1982.

83. W. D. Callister, Jr., *Materials Science and Engineering* pp. 98–103. Wiley: New York, 1985.

84. R. Manory, Effects of deposition parameters on structure and composition of reactively sputtered TiN_x films, *Surf Eng* 3(3), 233–238, 1987.

85. Z. Insepov, R. Manory, J. Matsuo, and I. Yamada, Proposal for a hardness measurement technique without indentor by gas-cluster-beam bombardment, *Phys Rev B* 61, 8744, 2000.

86. T. Michely and G. Comsa, Temperature dependence of the sputtering morphology of Pt(111), *Surf Sci* 256, 217, 1991.

87. C. Teichert, M. Hohage, T. Michely, and G. Comsa, Nuclei of the Pt(111) network reconstruction created by single ion impacts, *Phys Rev Lett* 72, 1682, 1994.

88. J. Krim, I. Heyvaert, C. Van Haesendonck, and Y. Bruynseraede, Scanning tunneling microscopy observation of self-affine fractal roughness in ion-bombarded film surfaces, *Phys Rev Lett* 70, 57, 1993.

89. S. Rusponi, C. Boragno, and U. Valbusa, Ripple structure on Ag(110) surface induced by ion sputtering, *Phys Rev Lett* 78, 2795, 1997.

90. M. Fried, L. Pogany, A. Manuaba, F. Paszti, and C. Hajdu, Experimental verification of the stress model for the wrinkling of ion-implanted layers, *Phys Rev B* 41, 3923, 1990.

91. T. M. Mayer, E. Chason, and A. J. Howard, Roughening instability and ion-induced viscous relaxation of SiO_2 surfaces, *J Appl Phys* 76, 1633, 1994.

92. E. Chason, T. M. Mayer, B. K. Kellerman, D. T. McIlroy, and A. J. Howard, Roughening instability and evolution of the Ge(001) surface during ion sputtering, *Phys Rev Lett* 72, 3040, 1994.

93. A. Gutzman, S. Klaumunzer, and P. Meier, Ion-beam-induced surface instability of glassy Fe40Ni40B20, *Phys Rev Lett* 74, 2256, 1995.

94. E. Snoeks, A. Pollman, and C. A. Volkert, Densification, anisotropic deformation, and plastic flow of SiO_2 during MeV heavy ion irradiation, *Appl Phys Lett* 65, 2487 1994.

95. R. M. Bradley and J. M. E. Harper, Theory of ripple topography induced by ion bombardment, *J Vac Sci Technol A* 6, 2390, 1988.

96. R. Cuerno, H. A. Makse, S. Tomassone, S. T. Harrington, and H. E. Stanley, Stochastic model for surface erosion via ion sputtering: Dynamical evolution from ripple morphology to rough morphology, *Phys Rev Lett* 75, 4464, 1995.

97. I. Koponen, M. Hautala, and O. -P. Slevanen, Simulations of ripple formation on ion-bombarded solid surfaces, *Phys Rev Lett* 78, 2612, 1997.

98. H. Kitani, N. Toyoda, J. Matsuo, and I. Yamada, Incident angle dependence of the sputtering effect of Ar-cluster-ion bombardment, *Nucl Instrum Methods B* 121, 489, 1997.

99. N. Toyoda, H. Kitani, N. Hagiwara, J. Matsu, and I. Yamada, Surface smoothing effects with reactive cluster ion beams, *Mater Chem Phys* 54, 106, 1998.

100. L. Cranberg, The initiation of electrical breakdown in vacuum, *J Appl Phys* 23, 518, 1952.

101. J. Knobloch, Advanced thermometry studies of superconducting radio-frequency cavities, PhD thesis, Cornell University, New York, 1997.

102. L. L. Laurent, High-gradient *rf* breakdown studies, PhD thesis, Stanford University, CA, 2002.

103. W. Dyke and J. K. Trolan, Field emission: Large current densities, space charge, and the vacuum arc, *Phys Rev* 89, 799, 1953.

104. W. P. Dyke, J. K. Trolan, E. E. Martin, and J. P. Barbour, *Phys Rev* 91, 1043, 1953.

105. H. H. Braun, S. Dobert, I. Wilson, and W. Wuensch, Frequency and temperature dependence of electrical breakdown at 21, 30, and 39 GH, *Phys Rev Lett* 90, 224801, 2003.

106. J. Norem et al., Dark current, breakdown, and magnetic field effects in a multicell, 805 MHz cavity, *Phys Rev ST Accel Beams* 6, 072001, 2003.

107. J. Norem, Z. Insepov, and I. Konkashbaev, Triggers for RF breakdown, *Nucl Instrum Methods A* 537, 510, 2005.

108. E. R. Buckle, A kinetic theory of cluster formation in the condensation of gases, *Trans Faraday Soc* 65, 1267–1288, 1969.

109. J. K. Lee, J. A. Barker, and F. F. Abraham, Theory and Monte Carlo simulation of physical clusters in the imperfect vapor, *J Chem Phys* 58, 3166–3180, 1973.

110. S. H. Bauer and D. J. Frurip, Homogeneous nucleation in metal vapors. 5. A self-consistent kinetic model, *J Phys Chem* 81, 1015, 1977.

111. P. A. Skovorodko, The peculiarities of condensation process in conical nozzle and in free jet behind it, *Proceedings of the 13th International Conferences on Rarefied Gas Dynamics*, p. 1053. Novosibirsk, 1982.

112. B. M. Smirnov, Model of clusterization in free jet expansion, *Phys Scripta* 50, 364–367, 1994.

113. B. M. Smirnov and A. Yu. Strizhev, Kinetics of clustering processes in expanding vapor, *Phys Scripta* 49(5), 615–618, 1994.

114. K. Yasuoka and M. Matsumoto, Molecular dynamics of homogeneous nucleation in the vapor phase, *J Chem Phys* 109, 8451–8462, 1998.

115. O. F. Hagena and W. Obert, Cluster formation in expanding supersonic jets: Effect of pressure, temperature, nozzle size, and test gas, *J Chem Phys* 56, 1793, 1972.

116. O. F. Hagena, Cluster beams from nozzle sources, in Peter P. Wegener (ed.), *Book Molecular Beams and Low Density Gasdynamics*, Chapter 2, pp. 93–181. Marcel Dekker: New York, 1974.

117. O. F. Hagena, Condensation in supersonic free jets, in L. Trilling and H. Y. Wachman (eds), *Rarefied Gas Dynamics*, vol. II, pp. 1465–1468. Academic, New York, 1969.

118. O. Abraham, J. H. Binn, B. G. DeBoer, and G. D. Stein, Gas dynamics of very small Laval nozzles, *Phys Fluids* 24(6), 1017–1031, 1981.

119. T. Takagi, I. Yamada, G. Takaoka, and H. Usui, *Proceedings of the Special Symposium on Advanced Materials-II*, p. 167, Osaka, 1990.

120. B. E. Wyslouzil, J. L. Cheung, G. Wilemski, and R. Stray, Small angle neutron scattering from nanodroplet aerosols, *Phys Rev Lett* 79, 431–434, 1997.

121. B. E. Wyslouzil, G. Wilemski, M. G. Beals, and M. B. Frish, Effect of carrier gas on condensation in a supersonic nozzle, *Phys Fluids* 6(8), 2845–2854, 1994.

122. B. E. Wyslouzil, C. H. Heath, J. L. Cheung, and G. Wilemski, Binary condensation in a supersonic nozzle, *J Chem Phys* 113, 7317–7329, 2000.

123. G. A. Bird, *Molecular Gas Dynamics and the Direct Simulation of Gas Flows*, Clarendon: Oxford, 1994.

124. T. Shimada and T. Abe, Applicability of the direct simulation Monte Carlo method in a body-fitted coordinate system, *AIAA J* 258–270, 1988.

125. T. Abe, Rarefied gas flow analysis by direct simulation Monte Carlo method in body-fitted coordinate system, *J Comput Phys* 83, 424–432, 1989.

126. S. D. Piersall and J. B. Anderson, Direct Monte Carlo simulation of chemical reaction systems: Simple bimolecular reactions, *J Chem Phys* 95, 971–978, 1991.

127. H. Hettema and J. S. McFeaters, The direct Monte Carlo method applied to the homogeneous nucleation problem, *J Chem Phys* 105, 2816, 1996.

CHAPTER 10

Crater Formation by Gas Cluster Ion Beam Impact

10.1 Introduction

This chapter addresses the following research fields: crater formations and their characterizations by SEM and AFM, surface sputtering and shock wave generation by gas cluster ion beams (GCIB) processing technology that has recently been developed in Japan for the semiconductor industry. Surface smoothing and modification, such as ripple or tip formation, will also be discussed. This chapter discusses the fundamental properties of cluster ion production, such as the design of supersonic nozzles, lateral sputtering, and nonlinearity effects in cluster ion beam sputtering. Further modifications to the GCIB method are proposed that would increase its smoothing capability. Several applications of the method were proposed, namely a new concept of absolute thin film hardness and mitigation of dark current in high-voltage accelerators by GCIB. The new method could produce the lowest achievable surface roughness, about 1–2 Å and works equally well on any type of surface: metallic, semiconducting, and insulating, including diamond films, and therefore GCIB can become a universal ultra-smoother device.

The structure of this chapter is as follows. In Section 10.3, crater formation by cluster ion impacts will be studied by hydrodynamics and molecular dynamics simulation results, as well as on experimental crater data obtained with large gas cluster ion impacts [1–3]. In Section 10.4, surface sputtering with ion beam irradiation will be discussed. In Section 10.5, surface sputtering effects will be explained based on the generation of microscopic shock waves by a large Ar gas cluster impact on a Cu(100) surface [4]. In Section 10.6, shock wave generation by a cluster ion impact will be discussed. A technique of sub-nano-hardness measurement that can be performed using cluster ion bombardment, and which should be free of substrate effects, will be discussed in Section 10.7 [4]. In Section 10.8, ripple and quantum dot structure formation will be addressed

[5,6]. In Section 1.7, the supersonic nozzle design will be reviewed based on direct simulation Monte Carlo simulation [7].

10.2   Hydrodynamic Theory of Crater Formation

The depth of a crater formed by cluster ion bombardment of a solid surface can be roughly estimated from the mass, momentum, and energy conservation laws assuming that the impact generates a shock wave:

$$E_0 = E_i + E_{hyd},\tag{10.1a}$$

$$E_i = \frac{1}{2}\bar{P}_H V,\tag{10.1b}$$

$$\bar{P}_H = P_c + P_{th},\tag{10.1c}$$

where:

E_0 is the total cluster ion energy

E_i and E_{hyd} are, respectively, the internal energy of a compressed area and the energy of a radial hydrodynamic motion of the compressed material encompassed by shock compression, as a whole

P_H is Hugoniot's pressure

V is the crater volume

P_c and P_{th} are the cold and thermal pressure components, respectively

For weak shock waves, with P_H well below 10^4 MPa, the two energy components on the right side of Equation 10.1a are equal, and P_{th} in Equation 10.1c can always be neglected in comparison with P_c. For example, at 30% compression the total pressure behind the shock wave for Pb has the following components: $P_c = 21.6$ GPa and $P_{th} = 3.35$ GPa [8]. According to this estimation, the internal thermal energy of the compressed material entering E_i can also be neglected, in comparison with the compression energy of cold material. (In fact, the target area adjacent to a crater may acquire enough thermal energy to be melted, and a rim around a crater can be formed by extrusion of the melt, due to plastic flow.)

While the radius of the hemispherical shock front is increasing in time, the mass of compressed target material increases proportional to the cube of the radius, which eventually reduces its energy.

The radius at which the shock wave stops could be estimated by equating Hugoniot's pressure, P_H, to the Brinell hardness number of the surface material. This means that at that radius, the shock cannot compress the material anymore. Taking into account the condition $E_i \sim E_{hyd}$, that gives the Brinell hardness value for the cold pressure from Equation 10.1a.

According to the formulas Equation 10.2, the crater depth should be proportional to 1/3 power of the total cluster energy: $h \sim E_0^{1/3}$. (This relation will not be valid for pressures above $\sim 10^5$ MPa, a case which is rare for cluster ion impacts with cluster energies below 300 keV; this estimate for the maximum attainable pressure could be obtained from the energy conservation law as $P_{max} < E_0/V_{cr}$, and use the crater radius ~ 100 Å from experiment [9,10].) Figure 10.1 shows the 1/3 power dependence of the crater depth on the total cluster energy calculated by MD for impacts of Ar_n ($n = 236, 370, 490$, and 736), with energies of 6.4–19.9 keV.

To examine the sensitivity of our MD crater depth results to a power law dependence exponent, two other dependencies are plotted in this figure: the short-dash line, with $\alpha = 1/4$, and the long-dash line, with $\alpha = 0.4$. As can be seen, the 1/3 power law is the best

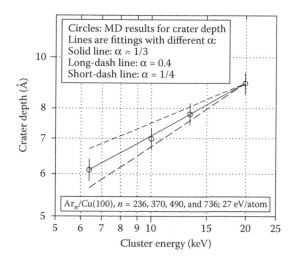

Figure 10.1 Crater depth calculated by molecular dynamics for Ar_n ($n = 236, 370, 490$, and 736) cluster ion impact with energy of 27 eV/atom on a Cu(100) surface (circles), the straight line corresponds to the $E^{1/3}$ power law. For comparison, the power-law dependencies with the exponents $\alpha = 0.4$ (long-dash line) and 0.25 (short-dash line) are also given.

fit of these results. An excellent linearity between the crater depths and the crater diameters was also obtained in our simulations.

The craters used for the data points in Figure 10.1 are shown in Figure 10.2 and were obtained by MD for Ar cluster impacts on a Cu surface, for the four cluster sizes given above which correspond to total cluster energies: 6.4 keV (a), 10 keV (b), 13.2 keV (c), and 19.9 keV (d).

Figure 10.3 shows measured values of crater diameters produced on gold with an Ar_n ion cluster beam of size $n=3000$, at increasing acceleration voltage. The dependence of the crater dimension on $E_0^{1/3}$ is evident. (The cluster energy E_0 is directly proportional to the acceleration voltage V_a.) The energy range suitable for crater formation experiments varies according to the nature of the cluster and the surface material but would typically be between 20 and 150 keV [9,10]. Hemispherical craters, with a linear dependence of the crater volume on the total cluster energy, are formed.

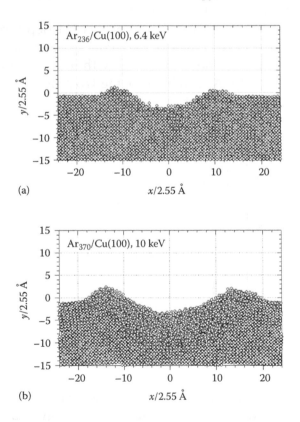

Figure 10.2   Side view of the craters formed on a Cu(100) surface calculated by MD for Ar_n ($n=236$, 370, 490, and 736) cluster ion impact, at 6.4 keV after 17 ps (a), and 10 keV (b).

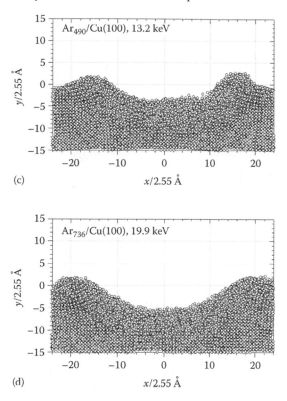

Figure 10.2   (*Continued*) Side view of the craters formed on a Cu(100) surface calculated by MD for Arn (*n* = 236, 370, 490, and 736) cluster ion impact, at 13.2 keV (c), 19.9 keV, and after 6 ps from the beginning of an impact (d).

The phenomena of cluster shock wave generation, crater formation, and surface sputtering are not possible at low cluster energies or for small cluster sizes. To form a crater on a surface, the cluster velocity should exceed several sound velocities of the surface material, which gives a cluster energy of about 20 eV/atom for Ar.

The cluster size should at least be of the order of the shock wave front thickness, which could be estimated to be approximately the same as in a planar shock wave, of the order of 50 Å [11]. Therefore, the crater formation phenomenon has thresholds in energy and in cluster size.

To complete the discussion on crater formation, it should be also mentioned that the energetic heavy monomer ion can also form a crater on a surface with a probability of about 1% [12], but the physics of crater formation is quite different than that of the cluster case [9,10]. Such craters, although rare, have occasionally been observed [12].

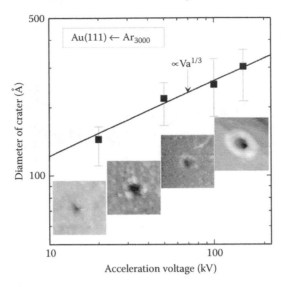

Figure 10.3   Energy dependence of the crater diameters obtained by STM (images of the craters are given below each data point) at Ar3000 cluster ion impacts, with total energy of 20–150 keV, on a Au(111) thin film surface deposited on mica. The straight line in this figure represents the 1/3 dependence of crater diameters on the acceleration voltage. (From D. Takeuchi, et al., *Nucl Instrum Methods Phys Res B*, 121, 493, 1997; T. Seki, et al., *Nucl Instrum Methods Phys Res B*, 121, 498, 1997.)

10.3   Simulation of Crater Formation by Energetic Cluster Ion Impact

MD calculations give positions and momenta of all cluster atoms and the target atoms in the central collision zone, which provide a wealth of information on the collision process and allow us to obtain a number of parameters of interest. Figure 10.4 shows a side view of the simulated crater formed by a 24 eV/atom Ar_{135} cluster impact on a Si(100) surface after a time interval of 14.4 ps from the start of cluster collision with the target. This picture shows a nearly triangular faceting of the crater, which is due to a higher binding energy of the (111) plane. The two dashed lines in Figure 10.4 show the directions of two (111) planes.

The top views of two craters on Si(100) and Si(111) surfaces are shown in Figure 10.5a,b. The figure shows a four-fold symmetry crater with facets formed by four (111) planes crossing the (100) surface. In Figure 10.5a, only those atomic positions are shown that lie below the previous unirradiated surface. Figure 10.5 shows a thin

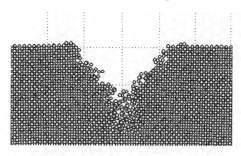

Figure 10.4   Side view of the crater formed by an Ar_{135} cluster impact, with energy of 24 eV/atom, on a Si(100) surface after 14.4 ps. (From L. P. Allen, et al., *J Appl Phys*, 92, 3671–3678, 2002.)

slice of the sample parallel to the surface that was made by cutting out the atoms in positions within the interval: -3 Å $< z_i < 0.05$ Å, where z_i is a Si atomic position coordinate normal to the surface.

The Si(111) surface shows faceting features quite different from that of the Si(100) surface as can be viewed in Figures 10.5b and 10.6c. We see that the side view (Figure 10.6c) has a round-shaped crater and the top view is also quite interesting. It has a six-point star shape, reflecting a hexagonal structure of the Si(111) lattice plane, not yet observed in simulation or in experiment. Figure 10.5b, a top

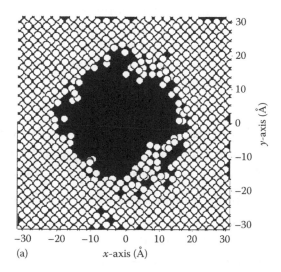

Figure 10.5   (a) Top views of the Si(100) surface showing the crater formed by an Ar_{135} cluster impact with energy of 24 eV/atom. (From L. P. Allen, et al., *J Appl Phys*, 92, 3671–3678, 2002; Z. Insepov, et al., *Nucl Instrum Methods B* 202, 261–268, 2003 [13]; Z. Insepov, et al., *Nucl Instrum Methods B*, 206, 846–850, 2003.) (continued)

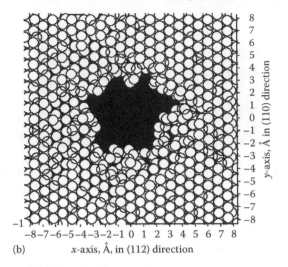

(b) x-axis, Å, in (112) direction

Figure 10.5 (*Continued*) (b) Top views of the Si(111) surface showing the crater formed by similar cluster size and energy. (From L. P. Allen, et al., *J Appl Phys*, 92, 3671–3678, 2002; Z. Insepov, et al., *Nucl Instrum Methods B 202*, 261–268, 2003; Z. Insepov, et al., *Nucl Instrum Methods B*, 206, 846–850, 2003.)

view for Si(111), shows a thin slice of the sample that contains four Si atomic layers parallel to the surface, with -4 Å $< z_i < 0.05$ Å.

The effect is very different from monomer ion irradiation and is clearly related to the different dynamics of collisions of the two projectiles with a solid. Cluster impact results in the transfer of kinetic energy to a large number of surface atoms, which leads to their displacement and local melting or sublimation.

10.4   Experiments with High-Resolution Transmission Electron Microscopy (HRTEM) and Atomic Force Microscopy (AFM)

10.4.1   High-resolution transmission electron microscopy

The images of individual gas cluster ion impacts were obtained using a JEOL 2010 HRTEM with a field-emission gun. Standard gluing, dimpling, and ion milling cross-sectional TEM sample preparation techniques were employed. Images were formed by orientation of the sample such that the transmitted beam was parallel to the <110> direction of the lattice and parallel to the (100) plane of the surface. An objective aperture that allows transmission of 13 beams was used to form the phase contrast images.

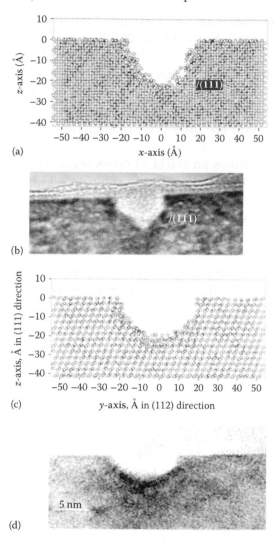

Figure 10.6 Comparison of simulated crater shapes with experiment. (a) Side view of the Si(100) surface showing a triangular-shaped crater created by a Ar_{135} cluster impact with an energy of 24 eV/atom. (b) HRTEM cross section image of individual 24 kV Ar gas cluster ion impact into Si(100). Crater conical edges align along the (111) lattice planes. (c) Side view of the Si(111) surface showing a round-shaped crater created by an Ar_{135} cluster impact with an energy of 24 eV/atom. (d) HRTEM cross-section image of an individual 24 kV O_2 gas cluster ion impact into Si(111). Rounder impact craters typically resulted from the O_2 cluster process, perhaps from the immediate oxidation of the Si and the change in associated bond strength of the forming crater wall. (From Z. Insepov, et al., *Nucl Instrum Methods B* 202, 261–268, 2003.)

Figure 10.6b shows an HRTEM cross section of an individual Ar gas cluster ion crater formed by a 24 kV acceleration into a Si(100) surface. The image cross-section shape agrees well with Figure 10.4, showing a conical impact crater with the (111) planes aligned with the crater walls. No subsurface dislocations are observed around the crater formed by the Ar gas cluster impact. The lattice integrity and orientation are preserved to the crater boundary and the surface.

Figure 10.6b shows an HRTEM cross section of a lower-energy 3 kV individual Ar gas cluster ion crater on a Si(100) surface. The shape and atomic plane cross-section boundaries appear identical to that of the 24 kV impact crater, but with a far shallower depth. In this cross section, a thin film of silicon crystal is imaged as intact over the impact crater. Such a thin film slice of silicon is envisioned for a cross-section image that is taken near but not exactly at the crater center. The material is thin enough to observe the back wall of the crater, with the outline of the full crater cross section observed.

Impact kinetics plays a first-order role in the GCIB to surface interaction, as observed by crater depth as a function of energy. This was again observed through the 3 kV individual impact crater images of an O_2 cluster into silicon as shown in Figure 10.5d. Once again, the general shape of the impact crater is maintained in the higher- and lower-energy processes, with the lower-energy impact having a shallower crater depth. In our observations, both the higher-energy and lower-energy O_2 gas cluster impacts typically result in a more rounded crater shape as compared with that of the Ar impact crater. This may be due to an inherently different cluster mass, charge, or an immediate silicon oxidation that, in turn, changes the bond strength that the remaining atoms of the impinging cluster "see" or will impact. No subsurface dislocations are observed around the crater formed by the O_2 gas cluster impact. As in the case for Ar gas cluster impacts, the lattice integrity is preserved. The factor of chemistry in GCIB processing is a topic of significant interest and recent publication [14].

10.4.2   Atomic force microscopy

In order to examine the impact crater morphology as compared with that of MD modeling, the surface of the GCIB processed silicon was also imaged by AFM. Prior to the individual cluster impacts into the silicon surface, the Si substrates exhibited ~2 nm of a native oxide as measured by spectroscopic ellipsometry (SE). Upon impact by the Ar gas cluster, the silicon from the substrate lattice (as well as the native oxide) is ejected or vaporized [14]. AFM imaging reveals hillock formation around and above the impact craters. The larger hillocks have been found to reflect the symmetry of the underlying Si substrate orientation.

Figure 10.7 shows an example of the result of the 24 kV Ar–GCIB individual impacts into Si surfaces with (100) and (111) substrate orientations. In the AFM image of Figure 10.7a, the symmetry of the Si

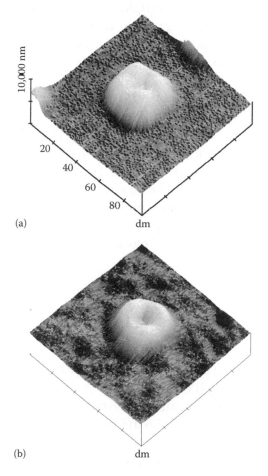

Figure 10.7 Three-dimensional 100 nm × 100 nm ($z = 10$ nm) AFM image of four-fold symmetry of Si(100) 24 kV Ar cluster impact hillock (a) and three-fold symmetry of Si(111) 24 kV Ar cluster impact hillock (b). (From L. P. Allen, et al., *J Appl Phys*, 92, 3671–3678, 2002; Z. Insepov, et al., *Nucl Instrum Methods B* 202, 261–268, 2003; Z. Insepov, et al., *Nucl Instrum Methods B*, 206, 846–850, 2003.)

substrate crystal is reflected by the distinct four-fold hillock formed by a 24 kV Ar gas cluster impact into the Si(100) surface. In the AFM image of Figure 10.7b, a three-fold symmetry can be observed in the hillock formed by a 24 kV Ar gas cluster impact into the Si(111) surface. The image axes are aligned with the substrate crystallographic directions, as shown. AFM images indicate that the larger hillocks are ~4 nm high, projecting above the native oxide, with a ~40 nm diameter. Smaller hillocks are numerous, without the clarity of symmetry found in the larger crater/hillock formations. The larger hillocks show distinct dimples in the center that correspond to the pits imaged by TEM. For low fluence, high-energy Ar cluster impacts, AFM has measured a continuum of individual hillock heights. They range from just over surface oxide height (~20 Å) to over 50 Å high.

Figures 10.4 through 10.6 indicate that simulation has good agreement with STM and AFM experiments, showing that the shapes of the craters depend on the surface crystallographic orientation. Craters on the Si(100) surface have a triangular pyramidal shape due to a higher density of the (111) plane and preferential crystallization of this plane. The location of the (111) plane is shown in Figure 10.1a,b as a white line. Craters on a Si(111) surface (Figure 10.6c,d) have a hemispherical shape. AFM images of the rims around the craters have a more complicated structure due to oxidation [1].

10.5   Surface Sputtering by Gas Cluster Ion Beam Irradiation

We use the MD model that we discussed in Chapter 2 in more detail for Si target sputtering [15–17].

Sputtering yields were derived from the saturation values of function $y(t)$, shown in Figure 2.6 in Chapter 2 of this book. They are shown in Figure 2.7 for cluster energies of 6.2, 10, 13.2 and 20 keV, and Ar cluster sizes of 236, 370, 490 and 736 atoms in a cluster, respectively. These results give the energy dependence of a sputtering yield: $Y \sim E^{\alpha}$, where E is the total cluster energy, and α is a parameter of nonlinearity.

As $E = \varepsilon N_{cl}$, where ε is the energy/atom and N_{cl} is the cluster size, we have $Y \sim \varepsilon^{\alpha} N_{cl}{}^{\alpha}$, where the cluster size enters explicitly. This formula facilitates comparison of the MD results obtained for a

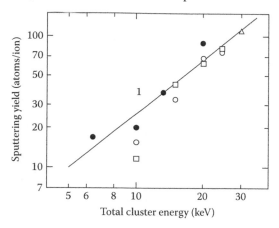

Figure 10.8 Calculated and experimental sputtering yields as a function of cluster energy. Line (1) shows fitting of the MD simulation data by a power law $Y \sim E^{1.5}$. Experimental data: O, Ar/Cu [18]; □, Ar/Ag [18]; Δ, Ar/Au [19]; ●, MD simulation results [20]. (Data from J. Matsuo, et al., *Nucl Instrum Methods B* 121, 459, 1997; I. Yamada, et al., *Nucl Instrum Methods Phys Res B* 82, 223, 1993; Z. Insepov, et al., *Mater Chem Phys* 54, 234–237, 1998.)

fixed energy/atom, or a fixed cluster velocity and variable cluster size with the experimental data obtained for a fixed cluster size and variable cluster energy defined by changing the accelerating voltage. This comparison is also shown in Figure 10.8.

Fitting the MD and the experimental results with the formula given above yields a value of the nonlinearity parameter $\alpha = 1.4$. This result is close to that attributed to the shock wave mechanism [21]. Another nonlinear sputtering model, the thermal spike model [22], gives a larger exponent than that obtained in the present paper. The figure shows that the MD results are reasonably close to the experimental data. The error in the experimental data for lower energy is larger, because of unreliable statistics in this case.

Figure 10.9 shows the angular dependence of a sputtered atoms for a 20 keV Ar cluster impact on a Cu surface. MD simulation results for different times after the impact are compared with experimental data points [23]. The maximum MD computation time of 10 ps was rather short to reveal a later stage of the sputtering yield, which is caused by thermal evaporation, oriented more in the normal direction. Qualitative agreement between the experiment and the calculations is nevertheless remarkable. We have suggested in an earlier work [24] that the lateral sputtering yield that occurred

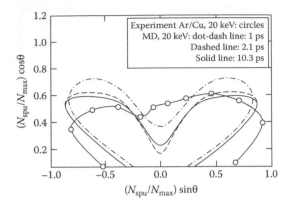

Figure 10.9   Angular distribution of sputtered atoms yield: dot-dashed line, 1 ps; dashed line, 2.1 ps; solid line, 10.3 ps; after the Ar_{236}, 20 keV, cluster impact obtained by MD, in comparison with experiment [23] shown by circles (○). (Data from N. Toyoda, et al., *Proceedings of the 14th International Conference on Application of Accelerators in Research and Industry*. Denton, TX, pp. 483–486, 1996.)

in an early stage of the cluster impact on a solid target may have a shock wave nature [25].

The main result of the lateral sputtering is the motion of highly energetic atoms along the surface. To support this idea, we have calculated the total flux of laterally moving atoms, due to cluster impact, as a function of time. The results are given in Figure 10.10

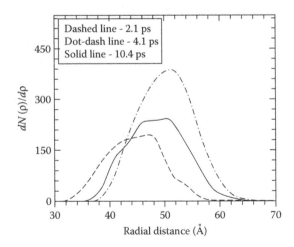

Figure 10.10   Radial displacement of ejected and redeposited Cu atoms by a 20 keV Ar_{736} cluster impact obtained in MD simulation. The abscissa is the radial distance measured from the center of a crater.

for three time instants 2.1, 4.1, and 10.4 ps after the 20 keV cluster impact on a Cu surface. As is seen from Figure 10.9, the redeposited atoms form a rim around the crater that evolves with time. The inner side of the rim moves slightly inside, partially filling the crater. Most of the atoms of the outer side of the rim are moving in an outward radial direction. As the surface kinetic energy is completely dissipated by 10 ps, we predict that the obtained crater shape will remain for a long time after the impact. Although the movement of the redeposited atoms is highly nonequilibrated, we can roughly estimate the surface diffusion coefficient from this figure to be of the order of 10^{-4} cm^2/s. This value is 10^6 times higher than the standard value for the Cu adatoms diffusion on Cu surfaces at room temperature [26].

10.6 Simulation of Shock Wave Generation

10.6.1 Simulation of 2D shock wave

The generation of shock waves by energetic gas cluster impact on a solid surface was studied by use of two- and three-dimensional molecular dynamics. The collisions of Arn ($n \sim 200$–350) clusters with the target of 40,000 atoms were modeled. The atomic scale shock waves arising from cluster impact have been obtained by calculating the pressure, temperature, and mass velocity of the target atoms. The asymptotic time dependence of distances traveled by a shock wave front is well described by a power law $R \sim t^\alpha$, with a constant $\alpha = 0.6$ for cluster energies between 17 and 85 eV/atom.

MD simulations of Ar$_{349}$ impacts, with energies of $E = 17$, 34 and 85 eV/atom on a hexagonal close packed (hcp) target were carried out. A corresponding 3D cluster would consist of around 3000 Ar atoms with total energies of 50, 100, and 250 keV, respectively. Figures 10.11 and 10.12 show that for cluster energies in the order of or greater than 50 keV shock waves are formed in the target. Figure 10.11 shows that the local pressure after the cluster impact has a roughly cylindrical shape with a very sharp abrupt front. The local target temperature calculated from tangential atomic degrees of freedom and depicted in Figure 10.12 reveals a time delay for the temperature front in comparison with the pressure front. This behavior distinguishes the cluster shock wave from the conventional one generated at macroscopic impacts where the pressure pulse rises together with the temperature pulse [8].

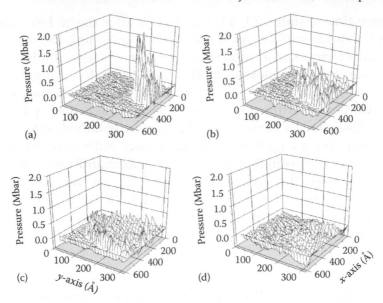

Figure 10.11   The local target pressure calculated from virial expression for four time instants of 362 fs (a), 724 fs (b), 1083 fs (c), and 2172 fs (d) after the Ar$_{349}$ cluster impact.

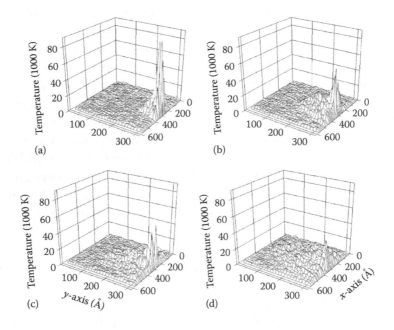

Figure 10.12   The local target temperatures calculated from tangential atomic velocities at the same conditions as in Figure 10.2.

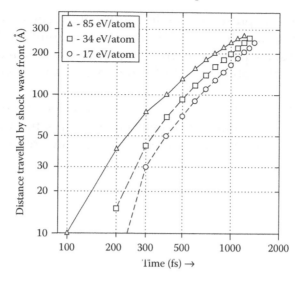

Figure 10.13 The shock wave trajectories calculated as distances traveled by the shock waves for the three energies at Ar_{265} cluster impact.

The shock wave trajectories depicted in Figure 10.13 were calculated from the distances traveled by the shock waves assuming cylindrical symmetry of a mass velocity field at a late impact stage. Fitting of these plots by the time dependence (1) shown in Figure 10.14 gives $\alpha = 0.6$. This figure corresponds to the shock wave velocity after an Ar_{265} cluster impact with 85 eV/atom. This result agrees well with the laser ablation experiment [25] where the planar shock waves were studied. A possible explanation of this agreement can be done if it is assumed that in the case of cluster impacts there are generated one-dimensional shock waves.

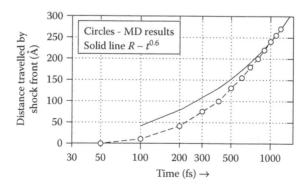

Figure 10.14 The fitting of the shock wave trajectory at Ar_{349} cluster impact, with the energy of 85 eV/atom, by a self-similarity law $R \sim t^{\alpha}$.

The pressure pulse shown in Figure 10.11 consists of separated patterns, which also supports this assumption. We have observed the same patterns of the compression waves in our three-dimensional calculations [24,27] for the case of low-energy cluster impacts.

In order to investigate the distribution of shock energy over different degrees of freedom, the kinetic energies of radial and tangential movement and the difference between them were calculated as

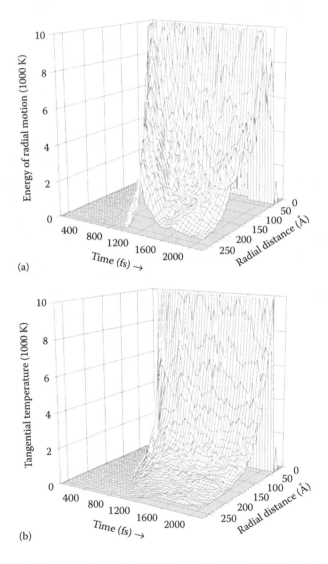

(a)

(b)

Figure 10.15   The energies of radial (a) and tangential (b) velocity components of the target's atoms.

plotted in Figure 10.15a,b, respectively. It can clearly be seen that thermal equilibrium exists in the area in front of the shock that has not yet been reached by the wave, whereas the vast area behind the front is in a nonequilibrium state in a time scale of several picoseconds. After a time interval in the order of 1–2.5 ps, depending on the impact energy, this region becomes equilibrated. The shock wave front remains under nonequilibrium condition for a longer time, as long as the impact energy is not dissipated. The same result has been revealed in [28], where the MD method was used to simulate a planar shock wave in solid Ar. This effect occurs due to the very short time of compression and heating, in the order of 10^{-13} s. The corresponding space scale is in the order of 10 Å.

Since the relaxation time over translational degrees of freedom is in the same order of magnitude, the overheating of radial degrees of freedom prevails against heating of tangential degrees. The pressure behind the shock front contains both the radial and the tangential components of temperature. Unfortunately, because of the non-stationary character of the cluster shock wave, it was not possible to calculate the radial component of temperature behind the shock front. We estimated the pressure assuming that the radial temperature is in the order of the tangential temperature. As is well known [8], local thermal equilibrium is an essential condition for correctly using hydrodynamic theory. A further item for the hydrodynamic approach is that the thickness of a shock wave front has to be much greater than the mean collisional free path. Both of these conditions are not accessible in the initial stage of cluster shock wave evolution.

In the latest stage of impact, as is evident from Figure 10.15, the thermal equilibrium can be attained. Therefore, hydrodynamics is useful for considering the latest stage of impact only. These results reveal a nontriviality of the atomic scale shocks generated by the accelerated cluster impact on a crystal surface.

The energy dependence of crater depth for Ar_{349} impacts with energies of 17, 34, and 85 eV/atom (solid circles in Figure 10.15), respectively, was fitted by a power law $E^{0.33}$ which is close to the prediction of self-similarity behavior [29].

10.6.2 Simulation of 3D shock wave

In various experimental STM observations [9,10] a hemispherical crater was obtained on the surface after cluster ion bombardment. Theoretical studies of heavy single ion impacts based on a shock wave viewpoint were performed based on a thermal diffusion equation or on Hugoniot's relation [21,30–33]. These works were successful in obtaining quantitative results regarding the physics of crater formation as well as estimates of sputtering yield, without dealing with the dynamics of shock waves or crater formation. As well, no relationship between ion energy and crater characteristics was obtained in these papers [21,30–33]. Large cluster ion impacts have been studied by a molecular dynamics method in [24,27,34–37]. MD has also been used to calculate the temperature, pressure, and energy of planar (one-dimensional) steady-state shock waves [11,28,38,39] to determine the velocity of a surface shock wave due to ion impact [40], to simulate a shock wave generation within a cluster [41], and to study cluster impacts [42].

Because of the inevitable non-steady-state character of the ion impact, it is very difficult to perform computer simulation of shock waves generated at such an impact. Webb and Harrison [40] were the first to calculate by MD the velocity of the shock wave generated with 5 keV Ar + ion impact on a Cu surface, to be 17.6 km/s. Hypervelocity Ar cluster impact on a rigid target surface and generation of a shock wave within a cluster have been modeled by MD in [41]. A two-dimensional (2D) MD method was used in Section 4.1, where shock wave generation was studied at an Ar cluster impact on a movable atomistic surface.

As we have shown, when a large gas cluster hits a solid surface with hypersonic velocity, it penetrates into the target as a whole to a depth that depends on cluster energy. A strong pressure wave of about 100 kbar is generated due to impact. In the present study, the dynamics of a hypervelocity Ar cluster impact on a Cu(001) surface is analyzed with a three-dimensional MD method. Clusters were formed from Ar atoms interacting via the Buckingham potential and an embedded atom method (EAM) potential was used to describe interactions between Cu atoms. The collisions of Ar_n ($n = 236-736$) clusters with a Cu(100) surface were modeled. The total number of target atoms was about 77,000 for energy of 6–13 keV, and about 10^5 for energy of about 20 keV. In our previous paper [16,42] a hybrid molecular dynamics (HMD) method was proposed, which combines conventional atomistic MD for the central cluster collisional zone with a continuum mechanics representation for the rest of the system.

This approach significantly reduces the system size and can keep the accuracy of the energy flow through the system boundaries.

According to this technique, the response of the continuum part to the atomistic MD part can be represented by two components—one that is determined by forces calculated from a stress tensor and depends on the magnitude of deformation of the boundary layers, and the second that controls the energy balance and is introduced by energy absorbing walls, which were simulated by thermal diffusion equations. In the present study the same boundary conditions were used as in [16] ("thermal" boundaries). The basic MD cell was divided into spherical layers of width dr and the local target variables such as temperature, pressure, energy, and the velocity of moving matter (mass velocity) within a spherical layer were calculated with a certain time step, for the whole computation time. Local target temperatures were obtained from the equipartition theorem by deducting atomic kinetic energies from the average kinetic energy for the given spherical layer, and local pressures were calculated from a virial formula [27,28,38,43].

A shock wave front in an ideal nonviscous and a nonthermal conductive gas is a zero-thickness surface that moves with hypersonic velocity. In a real solid it has a certain thickness defined by the real material viscosity and thermal conductivity [8]. At a shock front, the local temperature, pressure, and energy abruptly acquire an increase from their equilibrium values before the front, for example, room temperature and zero pressure, to much higher values behind the front. In a classical macroscopic shock, the pressure, volume or density, and temperature in front of and behind the wave are related through a simple formula known as Hugoniot's relation, which represents mass, momentum, and energy conservation laws [8].

The atomic scale shock wave emerging from the cluster impact was obtained as a steep increase of radial and transversal kinetic energies of the target atoms according to the technique described above for which a spherical layer thickness $dr = 3$ Å was used, as in [28,38]. The front of this rise was considered as a shock wave front. This definition of a shock wave front was used in [11,39] for a planar shock. The time and space dependence of radial kinetic energy for an Ar_{370} cluster impact with an energy of 27 eV/atom on a Cu(100) surface with a total cluster energy of 10 keV is shown in Figure 10.16a. Figure 10.16b shows the top view of Figure 10.16a with a cut of the z-axis at a certain level higher than room temperature, thus representing a trajectory of the pulse. The black areas in this figure correspond to temperatures higher than room temperature

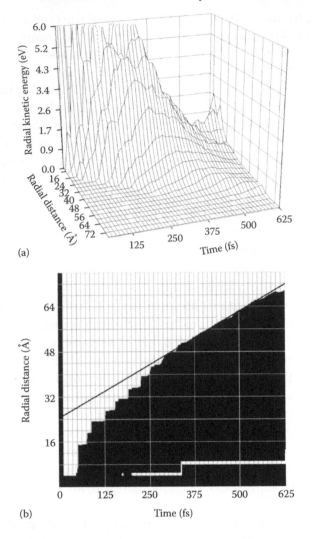

Figure 10.16   MD calculation of a shock wave generated at a Ar_{370} cluster ion impact on a Cu(100) surface, with energy of 27 eV atom (the total energy 10 keV). (a) Space and time dependence of kinetic energy within a spherical layer of 3 Å thickness at a radial distance r from the impact mark and at a time t from the beginning of the impact. (b) The bottom figure shows a top view of (a) with the z-axis cut at a temperature higher than room temperature thus showing a shock wave front trajectory. The black areas are the states behind the shock front, and the white areas in front of the shock.

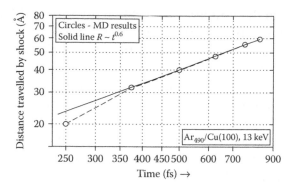

Figure 10.17   Distance traveled by shock front calculated by MD for Ar_{490} cluster ion impact with total energy of 13 keV on a Cu(100) surface (circles), the straight line corresponds to the $t^{0.6}$ power law.

(behind the shock wave front) and white areas show equilibrium states ahead of the front. As can be seen from this figure, a strong pulse, or shock wave, propagates into the solid. The velocity of the shock front could be obtained from this figure to be about 10 km/s. The shock wave penetrates to a distance of about 65 Å within a time interval of 0.5 ps, while the cluster itself penetrates to a lesser distance of about 10 Å.

Figure 10.17 shows that the velocity of the shock wave front rapidly decreases after about 0.6 ps and the shock wave almost disappears by this distance. Figure 10.17 also shows the radial distances $R(t)$ traveled by a shock front for a wave generated by an Ar_{490} cluster ion impact with total energy of 13 keV on a Cu(100) surface. Fitting this MD result by the time dependence $R(t) \sim t^\alpha$ gives $\alpha \approx 0.6$. The straight solid line in Figure 10.16b corresponds to this dependence. As can be seen from this figure, the dependence fits well for the latest time interval of the impact, after about 375 fs. The same value of α was obtained for impacts at 6, 10, and 19.9 keV. This time dependence of the distance traveled by a shock front could be easily measured experimentally. The calculated value is very close to the value 0.61 measured in a laser ablation experiment [25].

10.7   Simulation of Surface Smoothing

In simulating surface smoothing we have used the following length and time units for the numerical integration of the difference KS equation: $[L] = 1000$ Å, $[t] = 10^{-3}$ s. The x and y sides of the basic cell were equaled to the unit length, and that gives the cell's area equal to 1000×1000 Å². The height of the initial Gaussian hill was 100 Å. The ratio k/D_s in the second term of the KS equation for the Cu target has been estimated to be of the order of 10^{-17} cm²: $\gamma \sim 10^3$ dyn/cm, $\Omega \sim 10^{-24}$ cm², $n_0 \sim 10^{15}$ cm², and at room temperature.

We have computed the difference equation until the irradiation dose reached the experimental value, 5×10^{15} cm². Surface diffusivity, a critical parameter for describing this process, can be obtained from MD and is discussed in the next section.

In an earlier work we have shown that the sputtered atoms have a significant "lateral" momentum component [24]. A part of the ejected flux consisting of atoms, having lower normal momenta, can be redeposited on the surface. Some of these redeposited atoms have a high surface mobility which leads to a reduced surface roughness. Thus smoothing depends on the presence of energetic surface atoms emerging from the cluster impact site. This is a new nonequilibrium mechanism of surface modification that does not depend on heating or melting of the bulk solid, as would be expected in a usual surface diffusion mechanism.

The change in relative surface roughness as a function of irradiation dose is shown in Figure 10.18. Roughness is defined as root mean square (*rms*) j of the surface heights (Z coordinate) normalized

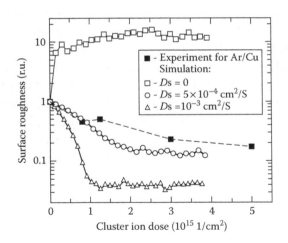

Figure 10.18   Surface roughness of Cu as a function of cluster ion dose. Experimental data for Ar on Cu are compared with calculations using different surface diffusivities. (Data from P. Sigmund and C. Claussen, *J Appl Phys*, 52, 990, 1981.)

to the initial value. The KS modeling with different surface diffusivity is compared with measurements of the roughness of the Cu surface after Ar cluster irradiation [44]. The figure shows that zero diffusivity leads to surface roughening, and the diffusion constant 5×10^{-4} cm²/s gives smoothing close to the experiment while 10^{-3} cm²/s gives a more rapid and larger smoothing effect. We propose that this effect is a result of the lateral motion of highly energetic atoms on the surface. To support this idea, we have calculated the total flux of laterally moving atoms, due to cluster impact, as a function of time. The results are given in Figure 10.10 for three time instants 2.1, 4.1, and 10.4 ps after the 20 keV cluster impact on a Cu surface. As is seen from Figure 10.10, the redeposited atoms form a rim around the crater that evolves with time. The inner side of the rim moves slightly inside, partially filling the crater. Most of the atoms on the outer side of the rim move in an outward radial direction. As the surface kinetic energy is completely dissipated by 10 ps, we predict that the obtained crater shape will remain for a long time after the impact. Although the movement of the redeposited atoms is highly nonequilibrated, we can roughly estimate the surface diffusion coefficient from this figure to be of the order of 10^{-4} cm² s⁻¹. This value is 10^6 times higher than the standard value for the Cu adatoms diffusion on Cu surfaces at room temperature [26].

10.8   Surface Hardness via Gas Cluster Surface Processing

At low and intermediate cluster beam energies, $E = 5$–100 keV, a hemispherical crater with a depth $h \sim E^{1/3}$ is created on a surface with a single cluster impact. As well, the sputtering rate when the surface is irradiated with many cluster ions is sufficiently high to be significant and measurable. The crater volume $V_{cr} \sim h^3$ has a linear dependence (2b) on the total energy divided by the material Brinell hardness

$$B \sim \frac{E_0}{V_{cr}}. \tag{10.2}$$

Thus, the crater depth h has a unique 1/3 dependence on the cluster energy and on the cold material Brinell hardness, a fact that was previously obtained for hypervelocity macroscopic body impacts on solid (mostly metal) surfaces.

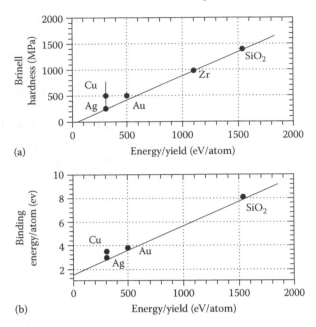

Figure 10.19   Brinell hardness data from available literature are correlated linearly to energy per sputtered atom (yield) (a), and binding energy
data are correlated with the energy per yield (b). Binding energy for SiO_2
8.12 eV is taken from Ya. B. Zel'dovich and Yu. P. Raizer, *Physics of Shock
Waves and High Temperature Hydrodynamic Phenomena*, Academic Press,
New York, 1967.)

Figures 10.19a,b present an analysis of available data which confirms these relationships. The correlation between the Brinell hardness number of the target material $-B$, and the energy per sputtered
surface atom is represented in Figure 10.19a and the correlation
between the binding energy of the target material per atom and the
energy per sputtered atom in Figure 10.19b. The sputtering yield
data were taken from [9]. The binding energy and surface hardness
data for gold, silver, copper, zirconium, and SiO_2 were found else-
where [45–47]. It is important to note here that SiO_2 was included
among the data points, which was otherwise obtained for metals.
SiO_2 is normally used for calibration of hardness measurement
equipment and the fact that the relationship for metals is also true
for SiO_2 should not be underestimated in this case, as it is a good
indication that our assumptions are correct.

As mentioned previously, crater volume and sputtering yield
should correlate with the binding energy of the material of a target.

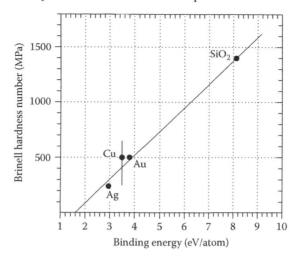

Figure 10.20 The feasible linear correlation between the Brinell hardness number, in MPa, and the binding energy, in eV/atom, of the target materials.

If that is true, a relationship could be obtained between B and energy per sputtered atom, B and binding energy, and between the energy per yield and binding energy. Figure 10.20 suggests that a linear correlation most probably exists between B and binding energy, but it should be noted here that surface hardness depends also on also on prior treatment of the metal. For example, hardening of a material due to a compressive load (work hardening) should lead to a different crater volume, and to a different sputtering yield. However, to our knowledge, the effect of strain hardening (cold working) on the surface binding energy has not been investigated, and therefore the proposed method should be suitable at this stage for "as-received" or annealed material only. These graphs contain only hardness data for material in its annealed form.

Regarding the stability of crater depth with energy and type of gas, this has been already shown, in both theory (Ch. 9, [24,27,34–37]) and experiment (Ch. 10, [9,10]), and these data are not shown again in the present study. Figure 10.21 shows the double-logarithmic linear dependence between the crater depth h and the hardness B where the total cluster ion energy is given as a parameter. The calibration was made for the calculated data obtained for Cu sputtering with Ar_{370} with energy 10 keV (solid line). The calibration of the lines was made for Cu as target material, for which there are available data for the Brinell hardness, sputtering yield, and binding energy. The

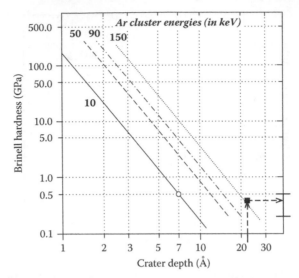

Figure 10.21   Brinell hardness data correlated with the crater depth, according to formula (10.2). (Data from F. A. McClintock and A. S. Argon, *Mechanical Behavior of Materials*, p. 455, MIT Press, Cambridge, MA, 1964 [48].)

circle corresponds to the crater depth of 7 Å obtained by the MD for 10 keV cluster impact and the Brinell hardness for Cu from Figures 10.19 and 10.20. The solid line is the calibration line obtained from one data point at 10 keV and with the slope from formula 10.2. The dashed line, dashed-dotted line, and the dotted line were obtained by shifting the solid line according to formula (6.3). The dashed arrows and the square show our prediction for surface hardness 300 MPa made by using the crater depth of 22 Å formed on a gold(111) surface by irradiation with a 150 keV Ar3000 cluster ion. The error bar on the right axis shows experimental hardness data from the literature.

The crater depth for this impact (cf. with Figure 10.2b) is about 7 Å and we can use the Cu hardness of 0.5 GPa from Figure 10.19 or 10.20. The line's slope was found from formula Equation 10.2. If we use this formula for different cluster energies, we can draw three other lines shown in this figure for 50 keV (dashed line), 90 keV (dash-dot line), and 150 keV (dot line). There is a scarcity of experimental data regarding crater dimensions as such measurements have rarely been performed. A crater depth of 22 Å was measured in [10] on a Au(100) surface bombarded with 150 keV Ar_{3000} cluster ions, and the Brinell hardness of gold is found from the graph as 300 MPa, which is within the error bar. More crater data need to be collected for materials with well-known BHN values so that the slope of the line and the correlation constant could be determined more accurately.

As $d \sim h \sim B^{-1/3}$, this obtaining can be used to calibrate various materials by the crater diameter (or depth), and in this way to define a new "true material hardness" scale that can be very useful, for example, for hard thin film coatings deposited on a soft substrate. It is proposed that this finding be used as a new technique for measuring surface hardness. The Brinell hardness B is in fact the cold material pressure P_c at which a crater is formed, so cluster ion impact is a tool for measuring the cold pressure curve, that is, it gives a "true hardness" of the material. This method does not use an indentor as the energy is instantaneously delivered by cluster impact. The damage occurs in the subsurface region and is not affected by the substrate. These elements lead to the conclusion that large gas cluster ions are potentially very suitable for use as hardness probes. The technique requires proper calibration, and a register of craters needs to be collected for future use as a hardness database. Thus, sputtering yield measurements might also be used as an alternative technique for measuring material hardness.

References

1. L. P. Allen, Z. Insepov, D. B. Fenner, C. Santeufemio, W. Brooks, K. S. Jones, and I. Yamada, Craters on silicon surfaces created by gas cluster ion impacts, *J Appl Phys* 92, 3671–3678, 2002.

2. Z. Insepov, L. P. Allen, C. Santeufemio, K. S. Jones, and I. Yamada, Computer modeling and electron microscopy of silicon surfaces irradiated by cluster ion impacts, *Nucl Instrum Methods B* 202, 261–268, 2002.

3. Z. Insepov, L. P. Allen, C. Santeufemio, K. S. Jones, and I. Yamada, Crater formation and sputtering by cluster impacts, *Nucl Instrum Methods B* 206, 846–850, 2003.

4. Z. Insepov, R. Manory, J. Matsuo, and I. Yamada, Proposal for a hardness measurement technique without indentor by gas-cluster-beam bombardment. *Phys Rev B* 61, 8744, 2000.

5. Z. Insepov and I. Yamada, Surface modification with ionised cluster beams: Modelling, *Nucl Instrum Methods Phys Res B* 148, 121–125, 1999.

6. Z. Insepov and I. Yamada, Surface processing with ionized cluster beams: Computer simulation, *Nucl Instrum Methods Phys Res B* 153, 199–208, 1999.

7. Z. Insepov and I. Yamada, Direct simulation Monte Carlo method for gas cluster ion beam technology, *Nucl Instrum Methods B* 202, 283–288, 2003.

8. Ya. B. Zel'dovich and Yu. P. Raizer, *Physics of Shock Waves and High Temperature Hydrodynamic Phenomena*, Academic Press: New York, 1967.

9. D. Takeuchi, K. Fukushima, J. Matsuo, and I. Yamada, Study of Ar cluster ion bombardment of a sapphire surface, *Nucl Instrum Methods Phys Res B* 121, 493, 1997.

10. T. Seki, T. Kaneko, D. Takeuchi, T. Aoki, J. Matsuo, Z. Insepov, and I. Yamada, STM observation of HOPG surfaces irradiated with Ar cluster ions, *Nucl Instrum Methods Phys Res B* 121, 498, 1997.

11. B. L. Holian, W. G. Hoover, B. Moran, and G. K. Straub, Shockwave structure via nonequilibrium molecular dynamics and Navier-Stokes continuum mechanics, *Phys Rev A* 22, 2798, 1980.

12. K. L. Merkle and W. Jager, Direct observation of spike effects in heavy-ion sputtering, *Philos Mag* 44, 741, 1981.

13. Z. Insepov, L. P. Allen, C. Santeufemio, K. S. Jones, and I. Yamada, Computer modeling and electron microscopy of silicon surfaces irradiated by cluster ion impacts, *Nucl Instrum Methods B* 202, 261–268, 2003.

14. I. Yamada, J. Matsuo, N. Toyoda, and A. Kirkpatrick, Materials processing by gas cluster ion beams, *Mater Sci Eng R* 34, 231, 2001.

15. Z. Insepov and I. Yamada, Computer simulation of crystal surface modification by accelerated cluster ion impacts, *Nucl Instrum Methods Phys Res B* 121, 44, 1997.

16. Z. Insepov, M. Sosnowski, and I. Yamada, Simulation of cluster impacts on silicon surface, *Nucl Instrum Methods Phys Res B* 127, 269–272, 1997.

17. Z. Insepov, M. Sosnowski, and I. Yamada, Surface smoothing with energetic cluster beams, *J Vac Sci Tech A* 15(3) (1997), pp. 981–984.

18. J. Matsuo, N. Toyoda, M. Akizuki, and I. Yamada, Sputtering of elemental metals by Ar cluster ions, *Nucl Instrum Methods B* 121, 459, 1997.

19. I. Yamada, W. L. Brown, J. A. Northby, and M. Sosnowski, Surface modification with gas cluster ion beams, *Nucl Instrum Methods Phys Res B* 82, 223, 1993.

20. Z. Insepov, I. Yamada, and M. Sosnowski, Sputtering and smoothing of metal surface with energetic gas cluster beams, *Mater Chem Phys* 54, 234–237, 1998.

21. Y. Kitazoe, N. Hiraoka, and Y. Yamamura, Hydrodynamical analysis of non-linear sputtering yields, *Surf Sci* 111, 381, 1981.

22. P. Sigmund and C. Clausen, Sputtering from elastic-collision spikes in heavy-ion-bombarded metals, *J Appl Phys* 52, 990, 1981.

23. N. Toyoda, J. Matsuo, and I. Yamada, The sputtering effects for cluster ion beams, *Proceedings of the 14th International Conference on Application of Accelerators in Research and Industry*. Denton, TX, pp. 483–486, 1996.

24. Z. Insepov, M. Sosnowski, and I. Yamada, Molecular dynamics simulation of metal surface sputtering by energetic rare-gas cluster impact, in I. Yamada et al. (ed.), *Laser and Ion Beam Modification of Materials*, p. 111, Elsevier, Amsterdam, 1994.

25. L. J. Dhareshwar, P. A. Paik, T. C. Kaushik, and H. C. Pant, Study of laser-driven shock wave propagation in Plexiglas targets, *Laser Part Beams* 10, 201, 1992.

26. H.-J. Ernst, F. Fabre, and J. Lapujoulade, Nucleation and diffusion of Cu adatoms on Cu (100): A helium-atom-beam scattering study, *Phys Rev B* 46, 1929, 1992.

27. H. Haberland, Z. Insepov, and M. Moseler, Molecular-dynamics simulation of thin-film growth by energetic cluster impact, *Phys Rev B* 51, 11061, 1995.

28. V. Yu. Klimenko and A. N. Dremin, Structure of a shock-wave front in a solid, *Sov Phys Dok* 25, 288, 1980.

29. R. Kinslow (ed.), *High-Velocity Impact Phenomena*, Academic Press: New York, 1970.

30. C. Ronchi, The nature of surface fission tracks in UO2, *J Appl Phys* 44, 3575, 1973.

31. G. Carter, Spike and shock processes in high energy deposition density atomic collision events in solids, *Radiat Eff Lett* 43, 193, 1979.

32. K. B. Winterbon, Shock waves in collision cascades, *Radiat Eff Lett Sect* 57, 89, 1980.

33. I. S. Bitensky and E. S. Parilis, Shock wave mechanism for cluster emission and organic molecule desorption under heavy ion bombardment, *Nucl Instrum Methods Phys Res B* 21, 26, 1987.

34. I. Yamada, J. Matsuo, Z. Insepov, and M. Akizuki, Surface modifications by gas cluster ion beams, *Nucl Instrum Methods Phys Res B* 106, 165, 1995.

35. T. Aoki, J. Matsuo, Z. Insepov, and I. Yamada, Molecular dynamics simulation of damage formation by cluster ion impact, *Nucl Instrum Methods Phys Res B* 121, 49, 1997.

36. I. Yamada and J. Matsuo, Gas cluster ion beam processing for ULSI fabrication, in advanced metallization for future ULSI, in K. N. Tu, J. W. Q. Mayer, J. M. Poate, and L. J. Chen (eds), p. 263, *MRS Symposia Proceedings* 427, Materials Research Society, Pittsburgh, 1996.

37. R. S. Averback and M. Ghali, MD studies of the interactions of low energy particles and clusters with surfaces, *Nucl Instrum Methods Phys Res B* 90, 191, 1994.

38. A. Paskin and G. J. Dienes, Molecular dynamic simulations of shock waves in a three-dimensional solid, *J Appl Phys* 43, 1605, 1972.

39. B. L. Holian, Shock waves and spallation by molecular dynamics, in M. Mareschal and B. L. Holian (eds), *Microscopic Simulations of Complex Hydrodynamic Phenomena*, Vol. 292 of NATO Advanced Study Institute, Series B: Physics, p. 75, Plenum: New York, 1992.

40. R. P. Webb and D. E. Harrison, Jr., Evidence for ion-induced hypersonic shock waves for computer simulations of argon ion bombardment of copper, *Appl Phys Lett* 39, 311, 1981.

41. U. Even, I. Shek, and J. Jortner, High-energy cluster: Surface collisions, *Chem Phys Lett* 202, 303, 1993.

42. Z. Insepov and I. Yamada, Molecular dynamics study of shock wave generation by cluster impact on solid targets, *Nucl Instrum Methods Phys Res B* 112, 16, 1996.

43. L. Verlet, Computer "experiments" on classical fluids. I. Thermodynamical properties of Lennard-Jones molecules, *Phys Rev A* 159, 98, 1967.

44. I. Yamada, J. Matsuo, Z. Insepov, D. Takeuchi, M. Akizuki, and N. Toyoda, Surface processing by gas cluster ion beams at the atomic (molecular) level, *J Vac Sci Technol* A14, 1996, 781–785.

45. R. Behrisch (ed.), *Sputtering by Particle Bombardment I, Physical Sputtering of Single-Element Solids*, Springer: Berlin, 1981.

46. J. Gspann, in P. Jena et al. (eds), *Physics and Chemistry of Finite Systems: From Clusters to Crystals*, Vol. 374 of NATO Advanced Study Institute, Series C: Mathematical and Physical Sciences, p. 1115, Kluwer, Dordrecht, 1992.

47. D. M. Teter and R. J. Hemley, *Science* 271, 53, 1996.

48. F. A. McClintock and A. S. Argon, *Mechanical Behavior of Materials*, MIT Press: Cambridge, MA, 1964.

Index

Printed and bound by CPI Group (UK) Ltd, Croydon, CR0 4YY

01/11/2024

01782621-0008